EMBOSS Administrator's Guide

The European Molecular Biology Open Software Suite (EMBOSS) is a high-quality package of open source software tools for molecular biology. It includes over 200 applications, primarily for molecular sequence analysis, integrated with a range of popular third-party software packages under a consistent and powerful command line interface. The tools are available from a wide range of graphical interfaces, including easy-to-use web interfaces and powerful workflow software.

- Tools are easily customised and configured for different purposes without recompilation.
- Tools are well suited for scripting and for web interfaces.
- Simple installation-wide configuration or multiple individual user-specific configurations.
- Environment variables allow the global program behaviour to be set and controlled conveniently.
- Database integration is flexible. A variety of local databases and access methods are supported and new ones are easily added.
- Data retrieval over the web is transparent. Remote servers are easily configured to provide the same type of access as from a local database.
- EMBOSS runs on almost every available UNIX platform, MS Windows XP, MS Vista, MS Windows 7 and Mac OS X.
- Regular updates and fixes are made available in a timely manner.

The *EMBOSS Administrator's Guide* is the official, definitive and comprehensive guide to EMBOSS installation and maintenance:

- Comprehensive and up-to-date – find all the information you need to configure, install and maintain EMBOSS, including recent additions for v6.2.
- Step-by-step instructions with real-world examples – saves you time and helps you avoid the pitfalls on all the common platforms.
- In-depth reference to database configuration – learn how to set up and use databases under EMBOSS
- Includes EMBOSS Frequently Asked Questions (FAQ) with answers – quickly find solutions to common problems.

Peter M. Rice is a Group Leader at the European Bioinformatics Institute (EMBL-EBI, Hinxton, UK), a centre for research and services in bioinformatics and part of the European Molecular Biology Laboratory (EMBL). Peter instigated EMBOSS in 1996 when he was based at the Sanger Centre (Hinxton, UK), with Alan Bleasby (SEQNET, Daresbury) and in collaboration with Thure Etzold (EMBL-EBI).

Alan J. Bleasby is a Senior Scientific Officer at EMBL-EBI. He developed the early EMBOSS programming library (AJAX) at Daresbury Laboratory (Warrington, UK) where he was responsible for the SEQNET UK national bioinformatics service. He now works full-time on EMBOSS.

Jon C. Ison has been a developer of EMBOSS since 2000. He too is based at EMBL-EBI and helps coordinate EMBOSS with Peter and Alan. He is currently working on ontologies, data integration and application development.

The EMBOSS team thanks the organisations who have generously supported the project, including the Wellcome Trust, the Medical Research Council, the Biotechnology and Biological Sciences Research Council (BBSRC) and EMBL-EBI.

EMBOSS
Administrator's Guide
Bioinformatics Software Management

Dr Alan Bleasby
EMBL European Bioinformatics Institute

Dr Jon Ison
EMBL European Bioinformatics Institute

Mr Peter Rice
EMBL European Bioinformatics Institute

CAMBRIDGE
UNIVERSITY PRESS

Shaftesbury Road, Cambridge CB2 8EA, United Kingdom

One Liberty Plaza, 20th Floor, New York, NY 10006, USA

477 Williamstown Road, Port Melbourne, VIC 3207, Australia

314–321, 3rd Floor, Plot 3, Splendor Forum, Jasola District Centre, New Delhi – 110025, India

103 Penang Road, #05–06/07, Visioncrest Commercial, Singapore 238467

Cambridge University Press is part of Cambridge University Press & Assessment, a department of the University of Cambridge.

We share the University's mission to contribute to society through the pursuit of education, learning and research at the highest international levels of excellence.

www.cambridge.org
Information on this title: www.cambridge.org/9780521188159

First published 2011

A catalogue record for this publication is available from the British Library

Library of Congress Cataloging-in-Publication data
Bleasby, Alan.
EMBOSS administrator's guide : bioinformatics software
management / Alan Bleasby, Jon Ison, Peter Rice.
     p. cm.
ISBN 978-0-521-18815-9 (pbk.)
1. Bioinformatics. 2. Computer programming – Management. 3. EMBOSS.
I. Ison, Jon, 1972– II. Rice, Peter, 1956– III. Title.
QH324.2.B595 2011
572.80285′53–dc22

                                                    2010051569

ISBN   978-0-521-18815-9   Paperback

Contents

Acknowledgements

EMBOSS acknowledgements

The EMBOSS developers would like to thank the Funding Bodies who have supported the project and the many people who have contributed. If you are omitted from the list below please accept our apologies and let us know. Special thanks to all our users who have given valuable suggestions, criticism and praise over the years.

Funding bodies

Biotechnology and Biological Sciences Research Council (BBSRC)

European Molecular Biology Laboratory – European Bioinformatics Institute (EMBL-EBI)

Medical Research Council (MRC)

Wellcome Trust

Contributors

Peter Rice and *Alan Bleasby* were the instigators of the project. Peter was the original project coordinator and, after working for Lion Bioscience and the Sanger Centre, moved to the EMBL-EBI. He runs the EBI Rice Group which houses the EMBOSS project.

Alan Bleasby wrote the original AJAX library while at Daresbury Laboratory and now coordinates the project with Peter Rice. After working at the MRC HGMP-RC, Alan moved to the EBI to work full time on EMBOSS in 2005.

Jon Ison has contributed to all areas of the project since 2000 and now helps coordinate the project working full time with Peter and Alan.

Peter, Jon and Alan wish to thank:

Michael Schuster for contributing code to interface with the ENSEMBL database.

Mahmut Uludag for various contributions.

Gary Williams for writing applications and documentation, providing user support and many other contributions.

Tim Carver, a long-standing contributor; he designed, implemented and supports **Jemboss**.

Lisa Mullan for many contributions at EMBOSS meetings. She wrote training materials, organised training courses and gave talks.

Ian Longden for work on all aspects of EMBOSS whilst working full-time on a Wellcome Trust grant and in particular, for incorporation of graphics.

David Martin for new applications, bug testing and documentation.

Guy Bottu for **wEMBOSS** documentation, many valuable bug reports and suggestions.

Nicolas Joly for many bug reports and suggestions.

Damian Counsell for contributions at EMBOSS meetings and work on the documentation.

Hugh Morgan for work on the graphics libraries.

Claude Beazley for work on CORBA integration.

Ranjeeva Ranasinghe, Waqas Awan and *Michael Hurley* for work on the protein structure applications.

Bijay Jassal for contributions at EMBOSS meetings and SRS interface support.

Val Curwen for contributing applications, documentation and developing training materials.

Richard Bruskiewich for work on GFF and windows.

Catherine Letondal for integration with PISE.

Kathryn Beal for integration with SPIN.

Thomas Laurent for work on the SRS interface.

Rodrigo Lopez for work on the CpG island applications and for general encouragement.

Sinead O'Leary for contributing applications.

Thon de Boer for ACD documentation

Mark Faller for work on **emma** and other projects.

Andre Blavier for the original windows port.

Martin Sarachu, who has sadly passed away, and *Marc Colet* for **wEMBOSS**.

Luke McCarthy for **EMBOSS Explorer**.

Peter Cock for FASTQ and other sequence format validation.

David Bauer, Thure Etzold, Martin Senger, Tom Oinn, Don Gilbert, Will Gilbert, Rodger Staden, Bill Pearson, Simon Kelley, Ewan Birney, Susan Jean Johns and anyone we've missed for their contributions.

Janet Thornton, Graham Cameron, Michael Ashburner, Martin Bishop and *Richard Durbin* for their support.

EMBOSS Administrator's Guide acknowledgements

The authors would like to thank the following individuals for contributions to the EMBOSS documentation or for production of the *EMBOSS Administrator's Guide*:

Katrina Halliday from Cambridge University Press for encouragement and suggestions.

David Martin for providing inspiration from a previous guide to EMBOSS installation.

Preface

Introduction to EMBOSS

The European Molecular Biology Open Software Suite (EMBOSS) is a high-quality, well-documented package of open source software tools for molecular biology. It includes over 200 applications for molecular sequence analysis and other common tasks in bioinformatics. It integrates the core applications with a range of popular third-party software packages under a consistent and powerful command line interface. The software has many useful features; for example, it automatically copes with data in a variety of formats and allows for transparent retrieval of sequence data from the web.

EMBOSS includes extensive C programming libraries with a clean and consistent API. There is much useful inbuilt functionality, for example the handling of the command line and common file formats, making it a powerful and convenient platform to develop and release bioinformatics programs. True to the spirit of open source, EMBOSS is free of charge to all and the code is licensed for use by everyone under the GNU General Public Licenses (GPL and LGPL). No one individual or institute 'owns' the code, or ever will. Under the terms of the licenses, it can be downloaded via the internet, copied, customised and passed on, so long as these same freedoms are preserved for others. Contributions are strongly encouraged!

EMBOSS is well established. It is used in demanding production environments reflecting the maturity of the code base. A major new stable version is released each year. For those who need the latest code, the current source code tree can be downloaded via CVS. There have been many thousands of downloads including site-wide installations by administrators across the world, catering for hundreds or even thousands of users. Many interfaces to EMBOSS are available including easy-to-use web interfaces and powerful workflow software, enabling applications to be combined into analysis pipelines.

Administration of EMBOSS

EMBOSS was conceived as an open platform for bioinformatics applications and their development after the libraries of the Wisconsin (GCG) package were made proprietary. It now supersedes that package in most areas. EMBOSS was designed to cater for professional bioinformaticians, who need simple, or at least well-documented, tooling that is easily interfaced with diverse systems. The programs are adaptable for use in different situations. AJAX command definition (ACD) files, written in plain text, define all application parameters and command line behaviour, such as default and permissible values, allowing applications to be customised and configured for different purposes without recompilation. EMBOSS has the benefit of freely accessible source code, so where requirements go beyond that which can be handled by changes to the ACD file, novel applications can be developed rapidly and at minimal overhead. There are no arbitrary limits on the amount of data that can be processed; the upper memory limit is determined by the available system memory.

The command line interface is powerful and consistent. Crucially, all user input, including command line parsing and user prompting, is handled automatically before the main application starts. This guarantees that applications receive correct values for all required options when running and will not reprompt the user for more information. This makes the tools particularly suitable for scripting and for web interfaces. EMBOSS configuration files allow installation-wide configuration and multiple individual user-specific configurations. Environment variables allow the global behaviour of the programs, including paths to data and other files, to be set and controlled conveniently.

Database integration in EMBOSS is flexible. EMBOSS supports a variety of local databases and access methods including flat and indexed files such as *EMBL* flatfiles and some BLAST indices. Local utilities and database systems may be defined as access methods and new databases and access methods are added easily. Retrieval of sequence data over the web is transparent and remote servers (e.g. SRS and MRS) may be defined as databases, often providing the same access to the user as from a local database.

EMBOSS runs on almost every available UNIX platform, MS Windows XP, MS Vista/7 and Mac OS X. The applications are reliable and will hold up in demanding, high-throughput environments. Nightly compilation tests are performed on a variety of platforms and quality assurance (QA) tests are run on all applications ensuring everything works as expected. Applications are tested for memory usage ensuring they do not leak or corrupt memory. Regular updates and fixes arising from these tests and from community bug reports are made available in a timely manner.

About the authors

Peter Rice

Peter Rice is a group leader at the European Bioinformatics Institute (EMBL-EBI, Hinxton, UK), a centre for research and services in bioinformatics and part of the European Molecular Biology Laboratory (EMBL). His group investigates and advises on the e-Science and Grid technology requirements of the EMBL-EBI, through application development plus participation in standards development. His group also houses the EMBOSS project. Peter instigated EMBOSS in 1996 when he was based at the Sanger Centre (Hinxton, UK), with Alan Bleasby (SEQNET, Daresbury) and in collaboration with Thure Etzold (EMBL-EBI). He left Sanger in 2000 to work for LION Bioscience, and in 2003 joined the EMBL-EBI.

Alan Bleasby

Alan Bleasby is a Senior Scientific Officer at EMBL-EBI. Alan developed the early EMBOSS programming library (AJAX) at Daresbury Laboratory (Warrington, UK) where he was responsible for the SEQNET UK national bioinformatics service. He moved to the UK Medical Research Council Human Genome Mapping Project Resource Centre (UK HGMP-RC) when the SEQNET and HGMP-RC services merged in early 1999, where he was Group Leader of the Proteomics Applications Group and coordinated EMBOSS. When the HGMP-RC closed in 2005, he moved to the EBI to work full-time on EMBOSS.

Jon Ison

Jon Ison is a Senior Scientific Officer at EMBL-EBI. He moved from the University of Leeds to the UK HGMP-RC in 1999 to work on the Collaborative Computing Project in Biosequence and Structure Analysis (CCP11), before taking the post of Software Specialist for the Proteomics Applications Group in 2000. He has been a lead contributor and developer of EMBOSS since then, moving in 2005 with Alan to the EMBL-EBI where he helps coordinate the project with Peter and Alan.

How to cite EMBOSS

Please cite EMBOSS where appropriate.

Rice P., Bleasby A. and Ison J. *The EMBOSS User's Guide*. Cambridge University Press.

Ison J., Rice P. and Bleasby A. *The EMBOSS Developer's Guide*. Cambridge University Press.

Bleasby A., Ison J. and Rice P. *The EMBOSS Administrator's Guide*. Cambridge University Press.

Rice P., Longden I. and Bleasby A. EMBOSS: The European Molecular Biology Open Software Suite. *Trends in Genetics* 2000 **16**(6):276–277.

The EMBOSS website. http://emboss.open-bio.org/

Conventions

EMBOSS *Guide* conventions

Command line sessions and commands

Examples of command line sessions or any other screen output look like this:

```
% seqret

Reads and writes (returns) sequences
Input (gapped) sequence(s): tembl:x65923
output sequence(s) [x65923.fasta]: stdout

>X65923 X65923.1 H.sapiens fau mRNA
ttcctctttctcgactccatcttcgcggtagctgggaccgccgttcagtcgccaatatgc
agctctttgtccgcgcccaggagctacacaccttcgaggtgaccggccaggaaacggtcg
cccagatcaaggctcatgtagcctcactggagggcattgccccggaagatcaagtcgtgc
tcctggcaggcgcgcccctggaggatgaggccactctgggccagtgcggggtggaggccc
tgactaccctggaagtagcaggccgcatgcttggaggtaaagttcatggttccctggccc
gtgctggaaaagtgagaggtcagactcctaaggtggccaaacaggagaagaagaagaaga
agacaggtcgggctaagcggcggatgcagtacaaccggcgctttgtcaacgttgtgccca
cctttggcaagaagaagggccccaatgccaactcttaagtcttttgtaattctggctttc
tctaataaaaaagccacttagttcagtcaaaaaaaaaa
```

In the above example the command, which is typed in on the command line, is given in **this format**. The text given in *this format* indicates a value that must be typed in or replaced.

Occasionally you will see commands and values referred to within the text. For example:

- **seqret -sbegin** *5* **-send** *25*

Program listings and code

Program listings will look like this:

```
#include "emboss.h"
/* @prog helloworld *********************************************
**
** Prints "Hello, World!" to the screen.
**
******************************************************************/
int main(int argc, char **argv)
{
        embInit("helloworld", argc, argv);

        ajFmtPrint("Hello, World!\n");

        embExit();
        return 0;
}
```

Occasionally, code is referred to within the text and it is given in `this format`. For example:

- The `main()` function above includes the function `ajFmtPrint`.

Other conventions

Software packages, where they are mentioned, appear in **this format**. For example:

- **HMMER** is an EMBASSY package that wraps third-party applications.

Applications, where they are mentioned, appear in **this format**. For example:

- **seqret** is an example of an EMBOSS program.

Options to programs appear like this:

- `-help`

Interfaces, where they are mentioned, appear in **this format**. For example:

- **Jemboss** is an example of an EMBOSS interface.

The name of a database, or parts of a database, appear in *this format*. For example:

- *EMBL* is an example of a sequence database.

Specific EMBOSS system files or directories appear in `this format`. For example:

- `.embossrc` and `emboss.default` are EMBOSS system files.

Environment variables, where mentioned, appear in `this format`. For example:

- `EMBOSS_ACDROOT` is an EMBOSS environment variable.

All other specific system items appear in `this format`. For example:

- A directory is an example of a system item, for example: `/home/auser/emboss/emboss/ajax/`

Special text blocks

Some special comments are offset from the main body of text.

Caution

A note of caution where there might be undesirable or unexpected consequences of some action.

Important

Comments or notes of special significance.

Note

Important but peripheral information to the main body of text.

Tip

Helpful hints, shortcuts, etc.

Welcome to the *EMBOSS Administrator's Guide*

Summary

This manual was written with newcomers to EMBOSS in mind. You will benefit from at least a basic appreciation of molecular biology and some familiarity with UNIX and the C programming language. You should know how to open, use, save and close files using a text editor. It will also help if you've used the EMBOSS programs and are familiar with the command line (see the *EMBOSS User's Guide*).

The EMBOSS website provides auxiliary and reference documentation not included in the printed version of the documentation, in particular complete documentation for the libraries and applications. If you cannot find what you need here, you should try searching the EMBOSS website (http://emboss.open-bio.org).

Chapter 1. Building EMBOSS

Complete instructions on how to download, build and configure the EMBOSS package. Information is given on compilation options, installation of the libraries and applications, post-installation setup and maintenance. Platform-specific installation notes and hints for troubleshooting an installation are included.

Chapter 2. Building EMBASSY

The EMBASSY packages include applications with the same look and feel as EMBOSS, but which are grouped together either for convenience, or because the applications are specialised, or they involve non-sequence-based analysis, or because different licensing conditions are in effect. Complete instructions on how to download, build and configure the EMBOSS packages are included.

Chapter 3. Building Jemboss

A guide to installing and setting up the **Jemboss** graphical user interface (GUI) to EMBOSS. Instructions are provided for installation as a standalone GUI, or in authenticating or non-authenticating client-server mode. Technical details such as authentication, the use of batch queuing software and configuration via the `jemboss.properties` file are explained. Platform-specific installation notes and hints for troubleshooting an installation are included.

Chapter 4. Databases

EMBOSS supports all the common data formats you are likely to need. This chapter gives comprehensive information on the configuration of databases and miscellaneous data files for use with EMBOSS. It explains the syntax used to write database definitions in the EMBOSS system file (`emboss.default`), including possible attributes and access methods. Ways to test your database definitions and the test databases included in EMBOSS are described. The indexing and configuring of flatfile, GCG format, BLAST, FASTA and other databases is covered, including ways to fine-tune a database installation and configure EMBOSS to use SRS for database lookup.

Appendix A. Resources

Summaries and links for software distributions that include EMBOSS, including RPM Package Manager files and various ports and packages.

Appendix B. Frequently Asked Questions (FAQ)

The list of EMBOSS frequently asked questions (FAQ) with answers.

Building EMBOSS

1

1.1 Downloading EMBOSS

1.1.1 Downloading with a web browser

You should use the URL:

```
ftp://emboss.open-bio.org/pub/EMBOSS/
```

The file you need for the EMBOSS base installation is

```
EMBOSS-latest.tar.gz
```

This will always be a link to the most recent version of EMBOSS. In the example given it is a link to EMBOSS-6.0.1.tar.gz. The other gz files are the EMBOSS-associated EMBASSY packages. These are extra files you can apply to the base EMBOSS installation. The EMBASSY packages are dealt with in a separate section of this book: you can download these optional files at the same time as EMBOSS-latest.tar.gz if you intend installing them.

1.1.2 Downloading by anonymous FTP

1.1.2.1 Interactive FTP

Change directory to the location in which you wish to download the source code. In this example you will download the source to /usr/local/src/EMBOSS. Then start your FTP client and point it to emboss.open-bio.org.

```
% ftp emboss.open-bio.org
Connected to emboss.open-bio.org (207.154.17.70).
220 (vsFTPd 2.0.1)
Name (emboss.open-bio.org:username):
```

The FTP server uses anonymous FTP so type in the username anonymous.

```
Name (emboss.open-bio.org:username):anonymous
     331 Please specify the password.
     Password:
```

Enter your *email address* here as the password for user anonymous. You could, in fact, type anything but it is a common courtesy to use your email address so the developers can get some idea of which sites have downloaded the software. Your email address will only be used for gathering such statistics. The FTP server will respond with something similar to the following:

```
230 Login successful.
Remote system type is UNIX.
Using binary mode to transfer files.
ftp>
```

You should now set your FTP client to use *passive mode*. How to do so depends on your FTP client. It is usually done by using the command **passive** or the command **pasv**; sometimes the FTP client has no such command and will usually already be in passive mode. To find out which, if any, command to type you can use the **help** command to show which commands your FTP client supports.

```
ftp> help
Commands may be abbreviated. Commands are:
!           cr          mdir        proxy       send
$           delete      mget        sendport    site
account     debug       mkdir       put         size
append      dir         mls         pwd         status
ascii       disconnect  mode        quit        struct
bell        form        modtime     quote       system
binary      get         mput        recv        sunique
bye         glob        newer       reget       tenex
case        hash        nmap        rstatus     trace
ccc         help        nlist       rhelp       type
cd          idle        ntrans      rename      user
cdup        image       open        reset       umask
chmod       lcd         passive     restart     verbose
clear       ls          private     rmdir       ?
close       macdef      prompt      runique
cprotect    mdelete     protect     safe
```

In this case the command needed is **passive**.

```
ftp>passive
Passive mode
```

If the server replies with *Passive mode off* then you were already in passive mode. In that case type **passive** again to make sure passive mode is on.

Now move to the directory containing the EMBOSS source code files and list the contents of that directory.

```
ftp> cd /pub/EMBOSS
250 Directory successfully changed.
ftp>ls
227 Entering Passive Mode (209,59,5,172,27,195)
150 Here comes the directory listing.
-rw-rw-r--    1   501  503     343812  Jul 15 19:45 CBSTOOLS-1.0.0.tar.gz
-rw-rw-r--    1   501  503     420983  Jul 15 19:45 DOMAINATRIX-0.1.0.tar.gz
-rw-rw-r--    1   501  503     462156  Jul 15 19:45 DOMALIGN-0.1.0.tar.gz
-rw-rw-r--    1   501  503     470537  Jul 15 19:45 DOMSEARCH-0.1.0.tar.gz
-rw-rw-r--    1   501  503   20204153  Jul 16 18:48 EMBOSS-6.0.1.tar.gz
lrwxrwxrwx    1   501  503         19  Jul 16 19:05 EMBOSS-latest.tar.gz -> EMBOSS-
6.0.1.tar.gz
-rw-rw-r--    1   501  503     390229  Jul 15 19:45 EMNU-1.05.tar.gz
-rw-rw-r--    1   501  503     431898  Jul 15 19:45 ESIM4-1.0.0.tar.gz
-rw-rw-r--    1   501  503     565686  Jul 15 19:45 HMMER-2.3.2.tar.gz
-rw-rw-r--    1   501  503     339939  Jul 15 19:45 IPRSCAN-4.3.1.tar.gz
drwxrwsr-x    7   501  503       4096  Feb 01 2006 Jemboss
-rw-rw-r--    1   501  503     450102  Jul 15 19:45 MEMENEW-0.1.0.tar.gz
-rw-rw-r--    1   501  503     365566  Jul 15 19:45 MIRA-2.8.2.tar.gz
-rw-rw-r--    1   501  503     445562  Jul 15 19:45 MSE-1.0.0.tar.gz
-rw-rw-r--    1   501  503     343305  Jul 15 19:45 MYEMBOSS-6.0.0.tar.gz
-rw-rw-r--    1   501  503     374766  Jul 15 19:45 MYEMBOSSDEMO-6.0.0.tar.gz
-rw-rw-r--    1   501  503    1624802  Jul 15 19:45 PHYLIPNEW-3.67.tar.gz
-rw-rw-r--    1   501  503     574940  Jul 15 19:45 SIGNATURE-0.1.0.tar.gz
-rw-rw-r--    1   501  503     532035  Jul 15 19:45 STRUCTURE-0.1.0.tar.gz
-rw-rw-r--    1   501  503     379929  Jul 15 19:45 TOPO-1.0.0.tar.gz
-rw-rw-r--    1   501  503     682200  Jul 15 19:45 VIENNA-1.7.2.tar.gz
drwxrwsr-x    3   522  503       4096  Aug 21 2006 contrib
drwxrwsr-x    2   501  503       4096  Nov 11 2005 doc
drwxrwsr-x    3   501  503       4096  Dec 15 10:41 fixes
drwxrwsr-x   10   501  503       4096  Jul 16 19:00 old
drwxrwsr-x    2   501  503       4096  Jul 06 2005 tutorials
drwxrwsr-x    3   501  503       4096  Jul 16 19:04 windows
226 Directory send OK.
ftp>
```

It is essential that you transfer these files as binary. The **help** command above shows that the command to achieve this is **binary**, so type it:

```
ftp>binary
200 Switching to Binary mode.
ftp>
```

Now download the source gz files.

```
ftp> get EMBOSS-latest.tar.gz
local: EMBOSS-latest.tar.gz remote: EMBOSS-latest.tar.gz
227  Entering Passive Mode (209,59,5,172,132,86)
150  Opening BINARY mode data connection for EMBOSS-latest.tar.gz (20204153 bytes).
226  File send OK.
20204153 bytes received in 67 seconds (2.4e+02 Kbytes/s)
ftp>
```

The file EMBOSS-latest.tar.gz is a link on the FTP server which points to the latest version of the EMBOSS source code. The directory listing from the **ls** command shows that, in this example, it points to EMBOSS-6.0.1.tar.gz. So, what you've really

3

downloaded is the EMBOSS 6.0.1 source code. The remaining gz files shown in the directory listing are the EMBASSY packages; these are EMBOSS-associated packages which you can optionally install once EMBOSS itself has been installed. If you intend installing the EMBASSY packages then now is a good time to get them too.

If you wish, you can download all the gz files using a single command as long as you set the server to turn prompting off. To do this use the **prompt** command:

```
ftp> prompt
Interactive mode off
ftp>
```

If, instead, the server responded with *Interactive mode on*, then the server already had prompting turned off; in that case type **prompt** again.

You can now download all the gz files using **mget *gz**. Note, however, that this will download both EMBOSS-latest.tar.gz and EMBOSS-6.0.1.tar.gz so there will be some unnecessary bandwidth used.

```
ftp> mget *gz
local: DOMAINATRIX-0.1.0.tar.gz remote: DOMAINATRIX-0.1.0.tar.gz
227  Entering Passive Mode (209,59,5,172,250,142)
150  Opening BINARY mode data connection for DOMAINATRIX-0.1.0.tar.gz
(349882 bytes).
226  File send OK.
349882 bytes received in 1.7 seconds (2e+02 Kbytes/s)
local: DOMALIGN-0.1.0.tar.gz remote: DOMALIGN-0.1.0.tar.gz
227  Entering Passive Mode (209,59,5,172,180,127)
150  Opening BINARY mode data connection for DOMALIGN-0.1.0.tar.gz
(347672 bytes).
.
. output truncated for clarity
.
ftp>
```

You can now exit from your FTP session with the command **quit**.

1.1.2.2   FTP using wget

The program **wget** can be used to download a remote directory non-interactively. More details on **wget** can be obtained from the Free Software Foundation (http://www.gnu.org). Assuming you have **wget** installed, use the following command; it will generate a lot of output on the screen:

```
% wget -m 'ftp://emboss.open-bio.org/pub/EMBOSS/'
--13:46:53--  ftp://emboss.open-bio.org/pub/EMBOSS/
           => 'emboss.open-bio.org/pub/EMBOSS/.listing'
Resolving emboss.open-bio.org... 207.154.17.70
Connecting to emboss.open-bio.org|207.154.17.70|:21... connected.
Logging in as anonymous ... Logged in!
==> SYST ... done.    ==> PWD ... done.
==> TYPE I ... done.  ==> CWD /pub/EMBOSS ... done.
```

```
==> PASV... done.    ==> LIST... done.

  [ <=>                              ] 1,501                --.--K/s

13:46:57 (171.74 KB/s) - 'emboss.open-bio.org/pub/EMBOSS/.listing'
saved [1501]

--13:46:57--  ftp://emboss.open-bio.org/pub/EMBOSS/DOMAINATRIX-
0.1.0.tar.gz
           => 'emboss.open-bio.org/pub/EMBOSS/DOMAINATRIX-0.1.0.tar.gz'
==> CWD /pub/EMBOSS... done.
==> PASV... done.      ==> RETR DOMAINATRIX-0.1.0.tar.gz... done.
Length: 349,882 (342K)
.
. output truncated for clarity
.
```

This command will have created a directory called emboss.open-bio.org/pub/EMBOSS and downloaded the gz files into that directory. Experienced UNIX users may take the opportunity to make a symbolic link called EMBOSS to this directory although it is not, of course, essential.

1.1.3   Unpacking the source code

You will have downloaded the EMBOSS source code to a suitable directory. Move to whichever directory you chose (e.g. cd /usr/local/src/emboss) and list the directory to make sure. We'll ignore the EMBASSY packages for now; they are described elsewhere in this book and, besides, you need to install EMBOSS before installing any EMBASSY package.

```
% ls
EMBOSS-latest.tar.gz
```

The EMBOSS-latest.tar.gz file is a compressed binary file and requires that your UNIX distribution has the **gunzip** program installed. Check that the command:

```
which gunzip
```

gives a positive response. If not then install the **gzip** package from whichever freeware site your UNIX distribution uses or, alternatively, compile the source code from the **gzip** homepage (http://www.gzip.org) which contains compilation instructions.

You unpack the EMBOSS distribution by typing:

```
gunzip EMBOSS-latest.tar.gz
```

This will create a file called EMBOSS-latest.tar. Such tar files are archive files containing the individual source code files.

You must now extract the archive using the **tar** program:

```
tar xf EMBOSS-latest.tar
```

This will create a new directory, EMBOSS-6.0.1: the exact name will depend on the version of EMBOSS being unpacked. Enter the directory and type **ls** to show the files. The directory listing should look something like this:

```
% cd EMBOSS-6.0.1
% ls
aclocal.m4
ajax
AUTHORS
ChangeLog
COMPAT
config.guess
config.sub
configure
configure.in
COPYING
depcomp
doc
emboss
.
. output truncated for clarity
.
```

If it doesn't then you're in the wrong directory. You are now ready to configure EMBOSS.

Caution

The **tar** program on most UNIX distributions will usually perform satisfactorily, however some will not. So, check the platform-specific notes (Section 1.7, 'EMBOSS installation: platform-specific concerns') to see whether you need to install the GNU version of the **tar** program instead.

1.2 Building EMBOSS

This comprises the following steps:

1. Configuration
2. Compilation
3. Installation
4. Post-installation.

It goes without saying that your system must have a C compiler installed before attempting to build the EMBOSS package. Many, if not most, operating systems come complete with a

C compiler. For those that don't (e.g. Solaris) you can install the GNU **gcc** compiler which is available from many freeware sites. EMBOSS has been optimized to use the standard/proprietary compiler for each of the operating systems to which the developers have access. So, for example, under IRIX the **cc** compiler should be preferentially used, under Linux distributions the **gcc** compiler will automatically be used.

1.3 Configuring EMBOSS

Configuring EMBOSS is done using the configure script provided with the package (see the directory listing Section 1.1, 'Downloading EMBOSS'). EMBOSS must be configured before it is compiled. There can be some pitfalls though, so look at the 'General prerequisites' (Section 1.7, 'EMBOSS installation: platform-specific concerns') for your operating system before continuing.

1.3.1 A simple configuration

At its most simple all you should need to type is:

```
./configure --prefix=/usr/local/emboss
```

The command above configures EMBOSS so that it will subsequently be installed into the directory /usr/local/emboss. Note that this is the final installation directory and not the directory containing the EMBOSS source code. The subsequent installation commands will then create subdirectories as necessary.

Installation is as follows:

- Binaries (/usr/local/emboss/bin)
- Shared libraries (/usr/local/emboss/lib)
- System-wide data (/usr/local/emboss/share/EMBOSS/data)
- Configuration files (ACD files) for the applications (/usr/local/emboss/share/EMBOSS/acd)
- Documentation (/usr/local/emboss/share/EMBOSS/doc)

To satisfy EMBOSS developers the source code header files are installed in: /usr/local/emboss/include.

The installation directory should, of course, be specified using a full path otherwise interesting failures may occur.

The output of the configuration will look something like this:

```
checking for a BSD-compatible install ... /usr/bin/install -c
checking whether build environment is sane ... yes
checking for gawk ... gawk
checking whether make sets $(MAKE) ... yes
checking for gawk ... (cached) gawk
checking for gcc ... gcc
checking for C compiler default output file name ... a.out
checking whether the C compiler works ... yes
```

```
checking whether we are cross compiling ... no
checking for suffix of executables ...
checking for suffix of object files ... o
checking whether we are using the GNU C compiler ... yes
checking whether gcc accepts -g ... yes
.
. output truncated for clarity
.
config.status: creating jemboss/resources/Makefile
config.status: creating jemboss/utils/Makefile
config.status: creating Makefile
config.status: executing depfiles commands
```

The installation process will automatically try to create your specified destination directory (e.g. /usr/local/emboss) if it doesn't already exist. However, if you do not have write permission in the parent directory then this will fail. So, at this stage, it is worthwhile to make sure that the installation directory can be created, e.g.:

```
mkdir /usr/local/emboss
```

and that you can create files in (have write access to) that directory. If the **mkdir** command above successfully creates the directory then you patently have write access to it; if not then log on as a system administrator, create the directory, and change the directory permissions accordingly using the **chown** and **chmod** UNIX commands.

We always recommend that you tell the configure script where you are going to install the compiled binaries by using the - -**prefix** option. If you don't use the - -**prefix** option then, like all such packages, EMBOSS will assume that it is to be installed into the /usr/local directory tree, thereby creating (if they don't already exist) the directories /usr/local/bin, /usr/local/lib, /usr/local/share, etc. This might not be a problem for you but remember that EMBOSS contains hundreds of programs and therefore sharing /usr/local/bin with other binaries you've put on the system may make it difficult for you to work out what else you've installed. It is your decision.

1.3.2 Configuring EMBOSS to use graphics

EMBOSS has the following graphic options as standard and there is usually no need to provide any special configuration options for them:

```
ps (postscript)
hpgl
hp7470
hp7580
meta
cps (colourps)
x11 (xwindows)
tekt (tektronics)
tek (tek4107t)
none (null, text)
```

8

```
data
xterm
svg
```

With the exception of x11 and xterm these are built into EMBOSS and require no external libraries. If you install the optional PNG graphics (see below) then two further options, png and gif, will appear in the above list.

It is theoretically possible to use the --without-x option of **configure** to specify that you do not want **X11** graphics. Though we appreciate that some sites may wish to provide EMBOSS on a non-X server we do not recommend disabling X. There is generally little system overhead benefit in doing so and this option has not been tested on all operating systems. We will address these issues in future EMBOSS releases.

The configuration will automatically successfully detect the location of **X11** on most systems and print out where it has been found. For example:

```
checking for X... libraries /usr/X11R6/lib64, headers /usr/X11R6/include
```

If you have installed **X11** in a completely non-standard place, or have multiple **X11** installations, then you may have to tell the configuration where you've put the one you want. This can be done using two options to **configure**:

```
--x-includes=DIR
--x-libraries=DIR
```

A common **X11** error for pre-6.1.0 versions of EMBOSS, when using the Linux operating system, would be that configuration and compilation would appear to progress satisfactorily but would eventually fail reporting that:

```
-1X11 cannot be found
```

This would happen if you hadn't installed the **X11** development packages (see Section 1.7, 'EMBOSS installation: platform-specific concerns'). EMBOSS releases from 6.1.0 onwards check more thoroughly for **X11** development files at the configure stage and are likely to terminate the configuration with a helpful message on failure to find them.

1.3.3   Configuring with PNG graphics

PNG graphics are an optional extra. PNG is, however, required if you intend using the **Jemboss** graphical interface to EMBOSS. It is also useful for producing output suitable for display in a web browser. The **configure** command will always attempt to find the files required to build in PNG graphics but most operating system distributions do not include the required files by default. If you require PNG graphics then you should read the section 'PNG prerequisites' (Section 1.7, 'EMBOSS installation: platform-specific concerns') for each operating system. You can tell the **configure** command where the support headers and libraries are using the - -**with-pngdriver=DIR** option. For example, if you have installed them under the directory root /usr/local/png then you would configure EMBOSS using:

```
./configure --prefix=/usr/local/emboss --with-pngdriver=/usr/local/png
```

For some operating systems, notably Linux, once the PNG support packages have been installed you do not need to specify the - -**with-pngdriver=DIR** option, but for most you will have to.

The configuration will report whether it has found all the PNG support files it needs. On success it will say:

```
checking if png driver is wanted ... yes
checking for inflateEnd in -lz ... yes
checking for png_destroy_read_struct in -lpng ... yes
checking for gdImageCreateFromPng in -lgd ... yes
PNG libraries found
```

If any of the -**lz**, -**lpng** and -**lgd** tests fail then it will report the failure. For example:

```
checking if png driver is wanted ... yes
checking for inflateEnd in -lz ... yes
checking for png_destroy_read_struct in -lpng ... yes
checking for gdImageCreateFromPng in -lgd ... no
No png driver will be made due to libraries missing/old.
```

Common PNG installation errors are:

- Failure to install all the PNG support files under the same directory root. The - - **with-pngdriver=DIR** option expects all PNG support files (including **ZLIB** and **gd**) to be under the same tree unless they are in standard UNIX system directories (the directories /usr/lib, /lib and /usr/include are examples of the latter).
- An error on running EMBOSS programs stating that there is a difference between the PNG version used in the compilation with that used when the program is run. This usually means that the **gd** library doesn't match the PNG library.
- You used a version of the **gd** library lower than v2.0.28. Versions below this did not contain both PNG and GIF support.

1.3.4 Configuring with PDF graphics

PDF graphics are an optional extra. The **configure** command will always attempt to find the files required to build in PDF graphics but most operating system distributions do not include the required files by default. If you require PDF graphics then you should read the section 'PDF prerequisites' (Section 1.7, 'EMBOSS installation: platform-specific concerns') for each operating system. You can tell the **configure** command where the support headers and libraries are using the - -**with-hpdf=DIR** option. For example, if you have installed them under the directory root /usr/local/pdf then you would configure EMBOSS using:

```
./configure --prefix=/usr/local/emboss --with-hpdf=/usr/local/pdf
```

For some operating systems, notably Linux, once the PDF support packages have been installed you do not need to specify the --with-hpdf=DIR option, but for most you will have to.

The configuration will report whether it has found all the PDF support files it needs. On success it will say:

```
checking whether to look for pdf support ... yes
checking for HPDF_New in -lhpdf ... yes
PDF support found
```

If the -lhpdf library is not found then the failure will be reported. For example:

```
checking whether to look for pdf support ... yes
checking for HPDF_New in -lhpdf ... no
No pdf support (libhpdf) found.
```

Note that the PDF support library may have a dependency on PNG subject to how it was built.

1.3.5   Configuration, Ensembl and SQL

The functionality of the Ensembl access library in EMBOSS is optional and depends entirely on whether or not you have installed code for either of the **MySQL** or **PostgreSQL** packages on your system, or indeed both. The EMBOSS configuration will automatically detect whether those SQL packages exist on your system and build Ensembl support appropriately. As long as either package is installed (or both) then Ensembl support will be enabled. Here is an example of SQL detection by EMBOSS:

```
checking for mysql_config ... /usr/bin/mysql_config
checking for MySQL libraries ... yes
checking for pg_config ... no
Not configuring for PostgreSQL
```

In this example **MySQL** has been successfully detected, but not **PostgreSQL**. So, in this case Ensembl support will be enabled and it will use the **MySQL** libraries.

Note that you will need to install the software development versions of the SQL packages. Many OS distributions will provide them pre-bundled with names such as:

- **mysql-devel**
- **postgreSQL-devel**

For other operating systems you will need to check for their availability and optionally install them as appropriate.

1.3.6   If you need to configure again

There are several reasons you may need to do this, not the least of which is if you realise you made a mistake in your original configuration. Another reason would be if your initial installation lacked PNG graphics support and you subsequently decide you need it.

   The first thing you should do, assuming your previous configuration ran to completion and you have already compiled EMBOSS, is to clean up the unpacked source code. You do this by typing the following in the same directory as the configure script:

```
make clean
```

The second thing you should do is not always necessary but it is advisable. It is to remove any information the previous **configure** command may have cached. Look, in the same directory as the **configure** command, for a directory called autom4te.cache. If this directory exists then delete it.

```
rm -rf autom4te.cache
```

If your operating system hasn't been updated for a few years it is possible that, instead of an autom4te.cache directory, there is a config.cache file. If so then delete it.

   It is now safe to configure EMBOSS again.

1.3.7   Advanced configuration options

The **configure** command has many other options; these can be shown using the --help option:

```
% ./configure --help

'configure' configures this package to adapt to many kinds of systems.

Usage: ./configure [OPTION] ... [VAR=VALUE] ...

To assign environment variables (e.g., CC, CFLAGS...), specify them as
VAR=VALUE. See below for descriptions of some of the useful variables.

Defaults for the options are specified in brackets.

Configuration:
-h, --help              display this help and exit
    --help=short        display options specific to this package
    --help=recursive    display the short help of all the included packages
-V, --version           display version information and exit
-q, --quiet, --silent   do not print 'checking...' messages
    --cache-file=FILE   cache test results in FILE [disabled]
```

```
-C, --config-cache  alias for '--cache-file=config.cache'
-n, --no-create     do not create output files
    --srcdir=DIR    find the sources in DIR [configure dir or '..']
```

```
Installation directories:
  --prefix=PREFIX           install architecture-independent files in PREFIX
                            [/usr/local]
  --exec-prefix=EPREFIX     install architecture-dependent files in EPREFIX
                            [PREFIX]
```

```
By default, 'make install' will install all the files in
'/usr/local/bin', '/usr/local/lib' etc. You can specify
an installation prefix other than '/usr/local' using '--prefix',
for instance '--prefix=$HOME'.
```

```
For better control, use the options below.
```

```
Fine tuning of the installation directories:
  --bindir=DIR           user executables [EPREFIX/bin]
  --sbindir=DIR          system admin executables [EPREFIX/sbin]
  --libexecdir=DIR       program executables [EPREFIX/libexec]
  --sysconfdir=DIR       read-only single-machine data [PREFIX/etc]
  --sharedstatedir=DIR   modifiable architecture-independent data [PREFIX/com]
  --localstatedir=DIR    modifiable single-machine data [PREFIX/var]
  --libdir=DIR           object code libraries [EPREFIX/lib]
  --includedir=DIR       C header files [PREFIX/include]
  --oldincludedir=DIR    C header files for non-gcc [/usr/include]
  --datarootdir=DIR      read-only arch.-independent data root [PREFIX/share]
  --datadir=DIR          read-only architecture-independent data [DATAROOTDIR]
  --infodir=DIR          info documentation [DATAROOTDIR/info]
  --localedir=DIR        locale-dependent data [DATAROOTDIR/locale]
  --mandir=DIR           man documentation [DATAROOTDIR/man]
  --docdir=DIR           documentation root [DATAROOTDIR/doc/PACKAGE]
  --htmldir=DIR          html documentation [DOCDIR]
  --dvidir=DIR           dvi documentation [DOCDIR]
  --pdfdir=DIR           pdf documentation [DOCDIR]
  --psdir=DIR            ps documentation [DOCDIR]
```

```
Program names:
--program-prefix=PREFIX            prepend PREFIX to installed program names
--program-suffix=SUFFIX            append SUFFIX to installed program names
--program-transform-name=PROGRAM  run sed PROGRAM on installed program names
```

```
X features:
--x-includes=DIR   X include files are in DIR
--x-libraries=DIR  X library files are in DIR
```

```
System types:
--build=BUILD   configure for building on BUILD [guessed]
--host=HOST     cross-compile to build programs to run on HOST [BUILD]
```

```
Optional Features:
--disable-option-checking ignore unrecognized --enable/--with options
--disable-FEATURE       do not include FEATURE (same as --enable-FEATURE=no)
--enable-FEATURE[=ARG]  include FEATURE [ARG=yes]
--disable-dependency-tracking speeds up one-time build
--enable-dependency-tracking do not reject slow dependency extractors
--enable-64             64 bit pointers
--enable-shared[=PKGS]  build shared libraries [default=yes]
--enable-static[=PKGS]  build static libraries [default=yes]
--enable-fast-install[=PKGS]
                        optimize for fast installation [default=yes]
--disable-libtool-lock  avoid locking (might break parallel builds)
```

```
--enable-debug              debug (-g option on compiler)
--enable-large              over 2Gb file support (default)
--enable-purify             purify
--enable-warnings           warnings (-Wall option on gcc compiler)
--enable-devwarnings        strict warnings (gcc compiler only) for developers
--enable-devextrawarnings      add extra warnings to devwarnings
--enable-mprobe             mprobe memory allocation test
--enable-savestats          save statistics and print with debug output

Optional Packages:
--with-PACKAGE[=ARG]        use PACKAGE [ARG=yes]
--without-PACKAGE           do not use PACKAGE (same as --with-PACKAGE=no)
--with-sgiabi               SGI compiler flags default=no
--with-gnu-ld               assume the C compiler uses GNU ld [default=no]
--with-pic                  try to use only PIC/non-PIC objects [default=use
                            both]
--with-tags[=TAGS]          include additional configurations [automatic]
--with-x                    use the X Window System
--with-docroot=DIR          root directory path of documentation defaults to none
--with-gccprofile           selects profiling
--with-java=DIR             root directory path of java installation
--without-java              to disable java
--with-javaos=DIR           root directory path of java installation include OS
--with-hpdf=DIR             root directory path of hpdf installation defaults to /usr
--without-hpdf              to disable pdf support
--with-pngdriver=DIR        root directory path of png/gd/zlib installation defaults to /usr
--without-pngdriver         to disable pngdriver usage completely
--with-auth=AUTHTYPE        defaults PAM
--with-thread=TYPE          thread type [default=linux]

Some influential environment variables:
  CC          C compiler command
  CFLAGS      C compiler flags
  LDFLAGS     linker flags, e.g. -L<lib dir> if you have libraries in a
              nonstandard directory <lib dir>
  LIBS        libraries to pass to the linker, e.g. -l<library>
  CPPFLAGS    C/C++/Objective C preprocessor flags, e.g. -I<include dir> if
              you have headers in a nonstandard directory <include dir>
  CPP         C preprocessor
  CXX         C++ compiler command
  CXXFLAGS    C++ compiler flags
  CXXCPP      C++ preprocessor
  F77         Fortran 77 compiler command
  FFLAGS      Fortran 77 compiler flags
  XMKMF       Path to xmkmf, Makefile generator for X Window System

Use these variables to override the choices made by 'configure' or to help
it to find libraries and programs with nonstandard names/locations.
```

Important

The configuration tools used by EMBOSS are those provided by the GNU project (http://www.gnu.org). The GNU tools can be regarded as a template which can be modified to enable successful building of a software suite. Such tailoring has been done by the EMBOSS package. As a result, many of the configuration commands are legacy ones which should not be used. We'll first say which options to avoid and then describe the EMBOSS-specific options which have a potential use

1.3.8   Configuration options to avoid

```
--exec-prefix
--bindir
--sbindir
--libexecdir
--datarootdir
--datadir
--sysconfdir
--sharedstatedir
--localstatedir
--libdir
--includedir
--oldincludedir
--infodir
--mandir
--srcdir
--with-docroot
--docdir
--htmldir
--dvidir
--pdfdir
--psdir
--program-prefix
--program-suffix
--program-transform-name
```

Use of any of the above options will have no effect or cause unpredictable results.

1.3.9   Configuration options to use

`--with-x, --without-x`	These options control whether **X11** support is enabled. By default `--with-x` is assumed and does not need to be added as an option. It's strongly recommend that **X11** support is not turned off. You gain nothing by doing so and this option has not been tested on many platforms.
`--x-includes=DIR` and `--x-libraries=DIR`	These options allow you to specify the location of your **X11** software if you've installed **X11** in a non-standard place. The EMBOSS configuration will usually correctly find the **X11** software so it should not be necessary to use these options.
`--disable-shared`	EMBOSS, by default, builds producing shared object libraries. Use of this option will force EMBOSS to build static libraries and executables. It will cause the compiled software to be significantly larger, but the executables then have no dependencies on any of the libraries. Using this flag may, under some circumstances, allow you to transfer executables between machines having similar operating systems. Most often, though, this

	option is used by EMBOSS developers to allow the use of software debugging tools.
--enable-warnings	This option will be ignored if your system does not use the GNU **gcc** compiler. If you are using **gcc** then several compilation warnings will be enabled. This option is used by developers as, with such a strongly-typed language as EMBOSS, all warnings are probably errors!
--enable-devwarnings	Turns on strict checking when used with the GNU **gcc** compiler. This flag is used by developers.
--enable-devextrawarnings	Turns on additional pedantic GNU **gcc** compiler warnings (e.g. padding alignment boundary checking) when used in conjunction with **--enable-devwarnings**. This flag is used by developers.
--with-gccprofile	This option should only ever be used with the GNU **gcc** compiler. It is used by developers to enable software profiling, i.e. it allows developers to locate potentially inefficient parts of their programs. The compiler profiling libraries must have been installed for this option to work.
--with-sgiabi	This option must be used if you are using the IRIX operating system. Its use is described in the platform-specific notes (Section 1.7, 'EMBOSS installation: platform-specific concerns') for IRIX.
--enable-64	This option is best avoided. Its only utility is for 32-bit systems where you want to force the use of 64-bit pointers. None of the EMBOSS programs require anything approaching 2 Gb of system memory. Also, the EMBOSS configuration by default will build optimally on 64-bit systems without this option – it has no utility on 64-bit systems.
--enable-debug	This option is provided for EMBOSS developers. It adds the appropriate compilation flag for using the GNU debugger **gdb**.
--with-pngdriver	This option has been described (Section 1.3.3, 'Configuring with PNG graphics').
--enable-large	This option gives EMBOSS the ability to access files greater than 2 Gb in size. This option is turned on by default and does not need to be specified. You can, if you wish, turn this ability off by using **--disable-large**.
--enable-purify	Sets flags for use with early versions of the PURIFY system. It is usually used in conjunction with the

--disable-shared option. In most cases you should ignore this option.

--enable-mprobe

This option is provided for EMBOSS developers. It turns on mprobe memory allocation checks.

--enable-savestats

This option is provided for EMBOSS debugging. It turns on statistical checks. It should not be used for a normal installation.

--with-java=DIR

Only for use if building EMBOSS with support for the **Jemboss** graphical user interface. This specifies the location of the **Java SDK**, e.g. /usr/local/jdk1.6.0_14/include. It is unlikely that you will ever need this option as the **Jemboss** installation script will automatically add it to the configuration command.

--with-javaos=DIR

Only for use if building EMBOSS with support for the **Jemboss** graphical user interface. This specifies the operating-system-specific location of the **Java SDK**, e.g. /usr/local/jdk1.6.0_14/include/linux. It is unlikely that you will ever need this option as the **Jemboss** installation script will automatically add it to the configuration command.

--with-auth=AUTHTYPE

Only for use if building EMBOSS with support for the **Jemboss** graphical user interface. This specifies the type of username/password protocol to be used by a **Jemboss** server. It is unlikely that you will ever need this option as the **Jemboss** installation script will automatically add it to the configuration command. The available AUTHTYPEs are:

- pam
- shadow
- rshadow
- noshadow
- aixshadow
- hpuxshadow

Where possible you should use pam authentication. shadow is for systems without re-entrant shadow password access functions, rshadow is for systems with re-entrant shadow password access functions, noshadow is for systems that can only access password information from /etc/passwd, aixshadow is for AIX operating systems and hpuxshadow is for HPUX operating systems.

Note

If using pam with Debian distributions then, on releases of EMBOSS prior to v6.1.0, before configuring do:

```
setenv CFLAGS "-DDEBIAN" or
CFLAGS="-DDEBIAN"
export CFLAGS
```

depending on which command shell you are using (the first for csh shells, the second for sh shells). This flag is automatically added by recent releases of EMBOSS.

`--with-thread=TYPE` Only for use if building EMBOSS with support for the **Jemboss** graphical user interface. This specifies the type of threading library to be used by a **Jemboss** server. It is unlikely that you will ever need this option as the **Jemboss** installation script will automatically add it to the configuration command. The possible values of `TYPE` are:

- linux
- freebsd
- solaris
- macos
- hpux
- irix
- aix
- osf

1.4   Compiling EMBOSS

After you have configured EMBOSS then the compilation stage consists of typing only one command:

```
make
```

This should be done in the same directory from which you invoked the **configure** command. Most operating systems come complete with a suitable **make** program. If the **make** command is not found then install it from your distribution DVD/CDs or install the GNU **make** application (recommended) from a freeware site or compile it from its source code (available from http://www.gnu.org/gnu/make/). Linux and Mac OS X distributions come complete with GNU **make**.

You will see hundreds of lines scroll up your screen as compilation progresses. The EMBOSS libraries (`plplot`, `ajax` (this is composed of several sub-libraries) and `nucleus`) will be compiled first, followed by the applications themselves. Compilation times will vary according to the power of your processor. On a 3 GHz processor it will typically take less than a handful of minutes.

If you have a multiprocessor machine then you can speed up the compilation by specifying how many processors to use, e.g.

```
make -j 4
```

will specify the use of four processors (or two processors with hyperthreading enabled).

After compilation has completed successfully you will see something similar to the following at the bottom of your screen.

```
..
Making all in utils
make[2]: Entering directory '/usr/local/src/EMBOSS-6.1.0/jemboss/utils'
make[2]: Nothing to be done for 'all'.
make[2]: Leaving directory '/usr/local/src/EMBOSS-6.1.0/jemboss/utils'
make[2]: Entering directory '/usr/local/src/EMBOSS-6.1.0/jemboss'
Not compiling Jemboss
make[2]: Leaving directory '/usr/local/src/EMBOSS-6.1.0/jemboss'
make[1]: Leaving directory '/usr/local/src/EMBOSS-6.1.0/jemboss'
make[1]: Entering directory '/usr/local/src/EMBOSS-6.1.0'
make[1]: Nothing to be done for 'all-am'.
make[1]: Leaving directory '/usr/local/src/EMBOSS-6.1.0'
%
```

1.4.1   Reporting compilation errors

If the compilation failed prematurely then recheck the platform-specific notes (Section 1.7, 'EMBOSS installation: platform-specific concerns') for your operating system to make sure you have all the prerequisites installed; then reconfigure and try again assuming you spotted a mistake. If all else fails then email emboss-bug@emboss.open-bio.org for help including:

- The output from the **configure** command
- The screen output from the **make** command (not just the last few lines)
- The config.status file
- The config.log file.

You can capture the output from (e.g.) the **make** command using one of:

```
make >&! make.out [csh shells]
make > make.out 2>&1 [sh shells]
```

Use the same method for reporting the output of the **configure** command.

1.5   Installing the libraries and applications

After having successfully configured and compiled the EMBOSS package then all the libraries and applications need installing to the directory tree you specified when you used the --prefix option of the **configure** command. This is done using one command:

```
make install
```

In fact, if the compilation had not already been done by typing **make** alone, then the above command would perform both the compilation and installation. We recommend, however, that you separate the two phases as we've described them; it helps in detecting where any failures may have happened.

Similarly to the compilation phase, many hundreds of lines will scroll up the screen. It is very rare for there to be errors in the installation phase as long as you checked that you had write permission in the directory tree you specified using --**prefix** in the configuration. If all has gone well then the last few lines of the installation output will look something like this.

```
make[3]: Leaving directory '/usr/local/src/EMBOSS-6.1.0/jemboss'
make[2]: Leaving directory '/usr/local/src/EMBOSS-6.1.0/jemboss'
make[1]: Leaving directory '/usr/local/src/EMBOSS-6.1.0/jemboss'
make[1]: Entering directory '/usr/local/src/EMBOSS-6.1.0'
make[2]: Entering directory '/usr/local/src/EMBOSS-6.1.0'
make[2]: Nothing to be done for 'install-exec-am'.
make[2]: Nothing to be done for 'install-data-am'.
make[2]: Leaving directory '/usr/local/src/EMBOSS-6.1.0'
make[1]: Leaving directory '/usr/local/src/EMBOSS-6.1.0'
%
```

Important

We recommend that you do not delete the EMBOSS source code tree. This is particularly important if you decided not to give a --**prefix** option to the **configure** command. For an explanation see Section 1.6.3, 'Deleting the EMBOSS package'.

1.6   Post-installation of EMBOSS

The most important post-installation step is to set your operating system environment so that it knows where to find the EMBOSS applications. Assuming that you followed our suggestion and configured EMBOSS using --**prefix=/usr/local/emboss** then you need to add the directory /usr/local/emboss/bin to your PATH. How to do this will depend on your operating system and the command shell you use. You can find out which shell you are using by typing:

```
env | grep SHELL
```

For users of the sh or bash shells (or derivatives) the PATH is altered using the following lines.

```
PATH="$PATH:/usr/local/emboss/bin"
export PATH
```

If you want to make these definitions available for all users then you would typically add the lines to the system /etc/profile file. If you just want to use EMBOSS yourself then you can add the lines to (e.g.) the .bashrc file in your home directory.

For users of csh or tcsh shells the PATH is altered using the following line.

```
set path= ($path /usr/local/emboss/bin)
```

If you want to make these definitions available for all users then you would typically add the lines to the system /etc/csh.cshrc file. If you just want to use EMBOSS yourself then you can add the line to (e.g.) the .cshrc file in your home directory.

Note

You may have to log out and log back in again for the changes to your **PATH** to take effect.

1.6.1 Testing the EMBOSS installation

An easy way to check that all is working is to use the EMBOSS application **embossversion**.

```
% embossversion
Writes the current EMBOSS version number
6.1.0
```

If the version number of EMBOSS is not printed similarly to the above then all is not well; if it is printed then celebrate appropriately.

1.6.1.1 Common errors during testing

The most common error is Command not found whenever you type in an EMBOSS application name. This is caused by incorrectly setting up the PATH (see above). Double-check that you set up the PATH correctly and, if necessary, take advice from someone familiar with the operating system you're using.

The second most common error is a report by the program that it cannot find the libnucleus library. This is one of the EMBOSS libraries and, if you followed our suggestion, it will be found in the /usr/local/emboss/lib directory after the installation phase. As long as you have set up your PATH correctly then EMBOSS should always be able to find its libraries. It has, however, been reported that some systems (notably SuSE Linux variants) have problems. In this case there are a few solutions.

1. With [Open]SuSE this error often happens if you have not specified a --**prefix** option or have otherwise installed EMBOSS at the root of the /usr/local directory tree such that the EMBOSS libraries are in the /usr/local/lib directory. [Open]SuSE maintains a cache of the contents of that directory which you will need to rebuild by typing as the superuser:

```
ldconfig
```

Do this also for other operating systems that maintain such a cache. If the error happens on other operating systems or distributions then you could do one of the following:

- Add the path to the EMBOSS libraries to the LD_LIBRARY_PATH environment variable. For example:

```
LD_LIBRARY_PATH="$LD_LIBRARY_PATH:/fu/bar/lib"
export LD_LIBRARY_PATH
```

- Or, for csh shells:

```
setenv LD_LIBRARY_PATH "$LD_LIBRARY_PATH:/fu/bar/lib"
```

2. Add the path to the EMBOSS libraries system-wide. This is perhaps the preferable way. For example, under Linux you could add the following line to the /etc/ld.so.conf file:

```
/fu/bar/lib
```

and then type:

```
ldconfig
```

For other operating systems, check the manual pages to see how to do the equivalent operations.

1.6.2   Post-installation of data files

If you wish to use the restriction mapping, domain recognition and amino acid index applications in EMBOSS then you will need to download the following databases from the internet; all are relatively small. Download them all to a temporary directory.

1.6.2.1   *REBASE*

This is available from ftp://ftp.neb.com/pub/rebase/

You need the `withrefm` and `proto` files from that directory. A common error is to download the `withref` file by mistake – it must be the `withrefm` file. The file extensions for these files change on the server every month to reflect the date.

Then type:

```
rebaseextract
```

and follow the prompts.

1.6.2.2   AAINDEX

This is available from ftp://ftp.genome.ad.jp/pub/db/community/aaindex/
You need the `aaindex1` file from that directory.
Then type:

```
aaindexextract
```

and follow the prompts.

1.6.2.3   PRINTS

This is available from ftp://ftp.ebi.ac.uk/pub/databases/prints/
You need the `prints.dat` file from that directory.
Then type:

```
printsextract
```

and follow the prompts.

1.6.2.4   PROSITE

This is available from ftp://ftp.ebi.ac.uk/pub/databases/prosite/release/
You need the `prosite.dat` and the `prosite.doc` files from that directory.
Then type:

```
prosextract
```

and follow the prompts.
You can now delete the data files you downloaded.

1.6.2.5   JASPAR

This is available from http://jaspar.genereg.net/html/DOWNLOAD/
You need the `Archive.zip` file. Uncompress it and then run:

```
jaspextract
```

and specify the all_data/FlatFileDir directory in response to the prompt. You can now delete the source directory contents.

1.6.3   Deleting the EMBOSS package

If you followed our advice and gave a - -**prefix** option to the **configure** command, thereby specifying a directory where EMBOSS alone would be installed, then there are two methods for deleting EMBOSS.

1. If you've kept the source code tree from which you'd done the **make install**. In this case, deleting the installation is easy. Just type:

```
make uninstall
```

This has the advantage that it will delete EMBOSS but will not delete any configuration files you have spent ages developing for your system. This is useful if you wish to reinstall a new version of EMBOSS after the deletion.

2. If you didn't keep the source code tree. As long as you specified a suitable - -**prefix** option to the **configure** command then you can use a UNIX **rm -rf directoryname** command to delete the EMBOSS installation tree.

If you didn't specify a - -**prefix** option to the **configure** command but did do a **make install** then you'll have to clean EMBOSS out of the /usr/local directory tree manually or, better, reinstall the same version of EMBOSS on top of itself and then use the **make uninstall** method.

1.6.4   Keeping EMBOSS up to date

From time to time, bug-fixes or new functionalities are provided, which can be applied to the version of EMBOSS you have installed. At such times new source code files will appear on our FTP server in the directory:

```
ftp://emboss.open-bio.org/pub/EMBOSS/fixes/
```

Usually these source code files are replacements for files that came with the EMBOSS distribution. You should read the README.fixes file in the above directory to see what the file fixes and whereabouts in the EMBOSS source directory it lives.

To apply the fixes, copy the source code file to its correct location, return to the top level of your EMBOSS source code tree and type:

```
make clean
make install
```

This is, of course, another very good reason for not deleting your EMBOSS source code tree.

A more convenient way of applying all the fixes from the above directory is to use the patch file in the sub-directory:

```
ftp://emboss.open-bio.org/pub/EMBOSS/fixes/patches/
```

The patch files are of the form `patch-1-n.gz` where n refers to the latest source code correction in the `README.fixes` file in the directory above. So, if there are ten corrections in the latter file then the patch file would be called `patch-1-10.gz`.

The patch files are applied using the UNIX **patch** command, e.g.:

```
gunzip EMBOSS-6.1.0.tar.gz
tar xf EMBOSS-6.1.0.tar
cd EMBOSS-6.1.0
gunzip -c patch-1-10.gz | patch -p1
```

Or, if the file has been uncompressed in transit:

```
patch -p1 < patch-1-10
```

You should always start with freshly extracted EMBOSS source code, as above, before applying a patch. This allows you to see any errors more easily. On rare occasions the developers will provide a patch file that contains fixes to a binary file. Some operating systems (e.g. FreeBSD) cannot handle binary patches and will report that such a patch file is malformed. In those circumstances follow the instructions in the `nonbinary` directory.

1.6.5 Installing a new version of EMBOSS

A new version of EMBOSS is released at least once per year, typically on St Swithun's Day (15 July). Before installing the new version you should either delete the existing EMBOSS version (if installing to the same directory) or install EMBOSS in a new location. Do not install a new version of EMBOSS on top of an existing installation as files from previous versions may cause compatibility problems.

Note

If you changed any system library or execution paths when you first installed EMBOSS then make sure you update these as necessary. A new version of EMBOSS is unlikely to work if new executables are trying to access older versions of the EMBOSS libraries.

1.6.6 EMBOSS configuration files

EMBOSS includes two files that are used to configure the package, particularly for defining databases and for making global settings that influence the behaviour of all EMBOSS programs.

The file emboss.default is used for site-wide configuration. Template files are included:

> Stable release (... **/share/EMBOSS/emboss.default.template**)
> CVS releases (... **/emboss/emboss/emboss.default.template**)

The file .embossrc, which you can create in your personal home directory, is used for user-specific customisation. Typically you might test, for example, database definitions in your own ~/.embossrc file before adding them to emboss.default.

1.6.6.1 Syntax of emboss.default and .embossrc Files

1.6.6.1.1 Blank lines and comments

Blank lines are ignored. Comments start with a '#' character in the first position of a line. For example:

```
# this is a comment
```

1.6.6.1.2 INCLUDE

INCLUDE allows you to include a subsidiary file as part of the text of the main emboss. default or .embossrc file at the position of the INCLUDE command. This is useful for keeping the configuration files tidy. For example, to include the contents of the file project_databases.def:

```
INCLUDE "project_databases.def"
```

1.6.6.1.3 Variable definitions

Variables may be set with the keyword SETENV (usually shortened to SET or ENV – either is OK), followed by the variable name, then the value to which you wish it set. For example:

```
SET dbdir /data/sequencedbs
```

This variable may now be used in the rest of the file emboss.default by preceding it with a $. For example:

```
file: $dbdir/data.dat
```

The name of the variable is case-insensitive when used within `emboss.default` or `.embossrc`.

1.6.6.2 Configuring EMBOSS for different groups of users

When maintaining EMBOSS for multiple users, more than one configuration might be required, for example to provide access to different sets of databases or data directories. It can be time-consuming and error-prone to maintain a series of individual `.embossrc` files in each user directory, or to force users to work in the same directory.

An alternative is to maintain one central copy of each of the different configuration files (`.embossrc`) in its own directory. All the user then need do is set the environment variable `EMBOSSRC` in their `.cshrc` (csh) or `.profile` (bash) file to point to the appropriate directory.

1.6.7   EMBOSS environment variables

Caution

It is possible to make EMBOSS unusable if you adjust the global variables. For example:

`SET EMBOSS_HELP 1`

will make all EMBOSS programs only display their help when they are run.

EMBOSS defines various environment variables. They include global variables used to control the behaviour of all EMBOSS programs, and variables to set the location of system files or directories, specify default values etc. There is normally no need to set the environment variables, but you may do so to customise the behaviour of your instance of EMBOSS.

Environment variables are useful for simplifying maintenance of your .embossrc file. If, for example, you specify the location of your databases as an environment variable, then if you move the databases you only have to update one line in the configuration file. For example, for the data directory:

```
/data/databases/flatfiles/
```

you might have something like this:

```
set EMBOSS_database_dir /data/databases/flatfiles
SET EMBOSS_embldir $EMBOSS_database_dir/embl
```

The second line sets another variable to the directory:

```
/data/databases/flatfiles/embl
```

Global environment variables must have UPPERCASE names and usually have boolean values; they can be turned on by setting them to "1", or "Y" (they are off by default.) The global variables can also be set in the UNIX session by defining an environment variable with the commands:

setenv *NAME* value (csh type shells)
export *NAME* =value (sh type shells)

where *NAME* is the name of the variable and value is the value you wish to set it to.

Table 1.1 Environment variables

Environment variable	Description	Type	Default value
EMBOSS_ACDCOMMANDLINELOG	Log file for full command line, used to convert quality assurance (QA) test definitions into memory leak test command lines	string	" "
EMBOSS_ACDFILENAME	Use filename rather than sequence name as default for file naming	boolean	N
EMBOSS_ACDLOG	Log ACD processing to file program.acdlog to debug ACD processing	boolean	N
EMBOSS_ACDPROMPTS	Number of times to prompt for a value interactively	integer	1
EMBOSS_ACDROOT	EMBOSS root directory for finding files	string	(install directory)
EMBOSS_ACDUTILROOT	EMBOSS source directory for finding files	string	(source directory)
EMBOSS_ACDWARNRANGE	Warn if a number is out of range and fixed to be within limits	boolean	N
EMBOSS_AUTO	Run with all default values unless -noauto is on the command line	boolean	N
EMBOSS_CACHESIZE	Cache size to use for database indexing	integer	2048
EMBOSS_DATA	EMBOSS directory for finding data files	string	(install directory)
EMBOSS_DEBUG	Write debug output to program.dbg unless -nodebug is on the command line	boolean	N

Table 1.1 (cont.)

Environment variable	Description	Type	Default value
EMBOSS_DEBUGBUFFER	Buffer debug output to save input/output time but risk losing output on a crash	boolean	N
EMBOSS_DIE	Print program abort messages to standard error	boolean	Y
EMBOSS_DOCROOT	EMBOSS directory for finding application documentation	string	*(install directory)*
EMBOSS_EMBOSSRC	Directory to search for an additional .embossrc file	string	*(current directory)*
EMBOSS_FEATWARN	Print warning messages when parsing feature table input	boolean	Y
EMBOSS_FILTER	By default read standard input and write to standard output unless -nofilter is on the command line	boolean	Y
EMBOSS_FORMAT	Input sequence format	string	unknown
EMBOSS_GRAPHICS	Default graphics output device	string	x11
EMBOSS_HOMERC	Read the .embossrc file in the user's home directory	boolean	Y
EMBOSS_HTTPVERSION	HTTP version	string	1.1
EMBOSS_LANGUAGE	(Obsolete) Language used for the codes.language file	string	english
EMBOSS_LOGFILE	System statistics log file	string	""
EMBOSS_MYEMBOSSACDROOT	**MYEMBOSS** package source directory for user's uninstalled utility ACD files	string	*(source directory)*
EMBOSS_NAMDEBUG	Write log nessages to standard error while processing .embossrc and emboss.defaults	string	N
EMBOSS_NAMVALID	Detailed validation while processing .embossrc and emboss.defaults	string	N
EMBOSS_OPTIONS	Prompt for optional command line values unless -nooptions is on the command line	boolean	N
EMBOSS_OUTDIRECTORY	Directory used to write output	string	*(current directory)*
EMBOSS_OUTFEATFORMAT	Output feature format	string	gff
EMBOSS_OUTFORMAT	Output sequence format	string	fasta

Table 1.1 (cont.)

Environment variable	Description	Type	Default value
EMBOSS_PAGER	Application to use for pages output to screen	string	more
EMBOSS_PAGESIZE	Page size to use for database indexing	integer	2048
EMBOSS_PROXY	HTTP proxy server address in the form `proxy.xyz.ac.uk:7890`	string	""
EMBOSS_RCHOME	Process the `.embossrc` file in the home directory	boolean	Y
EMBOSS_SEQWARN	Print warning messages when parsing standard sequence characters	boolean	N
EMBOSS_STDOUT	By default write to standard output unless `-nostdout` is on the command line	boolean	Y
EMBOSS_TIMETODAY	Date and time to override the current date – used to give a standard date and time for test runs	string	*2010–07–15 12:00:00*
EMBOSS_VERBOSE	Print verbose help output	boolean	N
EMBOSS_WARNOBSOLETE	Print warning messages when ACD file declares an application as 'obsolete'	boolean	Y

Table 1.2 Environment variables associated with global qualifiers

Environment variable	Description	Type	Default value
EMBOSS_ERROR	Print error messages to standard error	boolean	Y
EMBOSS_FATAL	Print fatal error messages to standard error	boolean	Y
EMBOSS_WARNING	Print warning messages to standard error	boolean	Y

1.6.7.1 Global qualifiers

EMBOSS includes several global qualifiers (see the *EMBOSS User's Guide*) that are available to all the applications. They are typically used by advanced users (who use **-options** or **-verbose**) or by developers (who use **-debug**, **-acdlog**). They may be set as follows:

Table 1.3 Environment variables to launch external applications

Environment variable	Description	Type	Default value
EMBOSS_CLUSTALW	Name or path to launch **clustalw**	string	clustalw
EMBOSS_PRIMER3_CORE	Name or path to launch **primer3_core**	string	primer3_core
EMBOSS_HMMALIGN	Name or path to launch **hmmalign**	string	hmmalign
EMBOSS_HMMBUILD	Name or path to launch **hmmbuild**	string	hmmbuild
EMBOSS_HMMCALIBRATE	Name or path to launch **hmmcalibrate**	string	hmmcalibrate
EMBOSS_HMMCONVERT	Name or path to launch **hmmconvert**	string	hmmconvert
EMBOSS_HMMEMIT	Name or path to launch **hmmemit**	string	hmmemit
EMBOSS_HMMFETCH	Name or path to launch **hmmfetch**	string	hmmfetch
EMBOSS_HMMINDEX	Name or path to launch **hmmindex**	string	hmmindex
EMBOSS_HMMPFAM	Name or path to launch **hmmpfam**	string	hmmpfam
EMBOSS_HMMSEARCH	Name or path to launch **hmmsearch**	string	hmmsearch
EMBOSS_MAST	Name or path to launch **mast**	string	mast
EMBOSS_MEME	Name or path to launch **meme**	string	meme
EMBOSS_MIRA	Name or path to launch **mira**	string	mira
EMBOSS_MIRAEST	Name or path to launch **miraEST**	string	miraEST
EMBOSS_BLASTPGP	Name or path to launch **blastpgp**	string	blastpgp
EMBOSS_FORMATDB	Name or path to launch **formatdb**	string	formatdb
EMBOSS_MODELFROMALIGN	Name or path to launch **modelfromalign**	string	modelfromalign
EMBOSS_NACCESS	Name or path to launch **naccess**	string	naccess
EMBOSS_RPSBLAST	Name or path to launch **rpsblast**	string	rpsblast
EMBOSS_STAMP	Name or path to launch **stamp**	string	stamp
EMBOSS_STRIDE	Name or path to launch **stride**	string	stride

```
set EMBOSS_QUALIFIER 1
```

where *QUALIFIER* is one of the global qualifiers. The value above is 1 but can be:

```
1 or Y for true
0 or N for false
```

Setting the qualifier value to true has the effect of running every program with that qualifier set. Qualifiers, when set, will work in the same way as if you used them when running the program. For example you can:

```
set EMBOSS_VERBOSE Y
```

and the program will run normally, but when the program is run with the `-help` qualifier, the output will be in verbose form.

Other program options that can be set include:

- EMBOSS_FORMAT
- EMBOSS_ACDROOT
- EMBOSS_DATA

The value of EMBOSS_FORMAT determines which default sequence format to use for output. For example, if you are running EMBOSS alongside GCG you may wish to have the following entry in your .embossrc:

```
set EMBOSS_FORMAT gcg
set EMBOSS_OUTFORMAT gcg
```

which has the effect of using GCG format for input and output by default.

If you wish to use a different directory for the ACD files then this can be set:

```
set EMBOSS_ACDROOT /path/to/acd
```

If you wish to maintain a separate data directory then use:

```
set EMBOSS_DATA /path/to/data
```

1.6.7.2 Logging

System administrators may wish to make use of the logging facilities of EMBOSS. Setting the variable EMBOSS_LOGFILE forces the system to keep a log of which programs are used when and by whom:

```
set EMBOSS_LOGFILE /site/log/emboss.log
```

The log file structure is very simple. Three tab-separated fields are stored, program name, user name, and the date and time:

```
prettyplot joeuser Wed Aug 02 14:29:13 2000
```

The file defined by EMBOSS_LOGFILE should be world writable. The following command ensures logging can occur:

```
chmod o+w /site/log/emboss.log
```

All settings can be overridden in a users .embossrc file by redefining the relevant variables. So to prevent your system usage being logged you can redefine EMBOSS_LOGFILE by putting the following entry in your .embossrc file:

```
set EMBOSS_LOGFILE /dev/null
```

This behaviour may change in the future to prevent users redefining some system settings.

1.6.7.3 Environment variables file (variables.standard)

Descriptions of the environment variables are stored in the EMBOSS system file variables.standard which is stored and installed in the application ACD file directory. An excerpt of this file is shown below:

```
acdcommandlinelog string "" "Log file for full commandline, used to convert
QA test definitions into memory leak test command lines"
acdlog boolean "N" "Log ACD processing to file program.acdlog"
acdprompts integer "1" "Number of times to prompt for a value when interactive"
acdroot string "(install directory)" "EMBOSS root directory for finding files"
acdutilroot string "(source directory)" "EMBOSS source directory for finding files"
```

1.6.8 EMBOSS data files

EMBOSS data files are included in the distribution and stored in the standard EMBOSS data directory, which can be defined by the EMBOSS environment variable EMBOSS_DATA.

If you built EMBOSS using **make install**, EMBOSS will by default install the data files, including those installed with **rebaseextract**, **prosextract**, **printsextract**, **aaindexextract** or **cutgextract**, in the directory:

```
share/EMBOSS/data
```

under the install directory, which is defined by the **--prefix** when you configured the package. Typically this is:

```
usr/local/emboss/share/EMBOSS/data.
```

If EMBOSS was not installed using **make install** but just compiled using **make**, then by default the data files are in:

```
emboss/data
```

under the directory where emboss was built.

If you want to keep your data files somewhere else, or have a set of data files you want to keep separate from those distributed with the package, then you can set the EMBOSS_DATA environment variable in your emboss.default or .embossrc file.

To see the available EMBOSS data files, run:

```
embossdata -showall
```

To fetch one of the data files into your current directory for you to inspect or modify, run:

```
embossdata -fetch -file EDatafileName.dat
```

where EDatafileName.dat is the name of the data file.

Users can provide their own data files in their own directories. Project specific files can be put in the current directory or, for tidier directory listings , in a sub-directory called ".embossdata". Similarly, for files to be accessible to all EMBOSS applications, invoked from any location, they can be put in your home directory, or in a sub-directory under it called ".embossdata".

The directories are searched in the following order:

- * . (your current directory)
- * .embossdata (under your current directory)
- * ~/ (your home directory)
- * ~/.embossdata

1.7 EMBOSS installation: platform-specific concerns

1.7.1 Linux RPM distributions

These distributions include Fedora, [Open]SuSE, Mandriva, CentOS and RedHat.

1.7.1.1 General prerequisites

Make sure you have the **X11** development files installed before you configure EMBOSS. The latest Linux distributions use the XORG version of **X11**. For these type one of:

```
rpm -q xorg-x11-proto-devel
rpm -q xorg-x11-devel
```

to check whether the package is installed. If not then install the relevant RPM from your distribution DVD/CDs using one of:

```
rpm -i filename.rpm
yum install packagename (if available)
```

If you wished to see, for example, all the `xorg-x11` packages installed on your system you would type:

```
rpm -qa "xorg-x11*"
```

Older Linux distributions use the **XFree86** version of **X11**. For these, type:

```
rpm -q XFree86-devel
```

to see whether the development files are installed.

1.7.1.2 PNG prerequisites

PNG support is optional; however, it is required if you intend using the **Jemboss** graphical user interface. You need to make sure that the following libraries and development files have been installed from your distribution DVD/CDs:

- **zlib**
- **zlib-devel**
- **libpng**
- **libpng-devel**

35

- **gd**
- **gd-devel**

You can check to see they're there using the **rpm -q** or **rpm -qa** command(s) as above. Use the **rpm -i filename.rpm** command to install them if necessary (or **yum install packagename** or a graphical package manager if available).

1.7.1.3   PDF prerequisites

PDF support is optional. You need to make sure that the following library and its development files have been installed from your distribution DVD/CDs:

- **libharu**
- **libharu-devel**

You can check to see they're there using the **rpm -q** or **rpm -qa** command(s) as above. Use the **rpm -i filename.rpm** command to install them if necessary (or **yum install packagename** or a graphical package manager if available).

1.7.1.4   Java

Java is optional; however, it is required if you intend using the **Jemboss** graphical interface. Many Linux distributions contain either outdated versions of Java or support only via the GNU **gcj** compiler. Using the former is not recommended: using the latter will not work at all! Some current Linux distributions are supplied with the **OpenJDK** package. This should work with **Jemboss** and we will aim to sort out any incompatibilities should they arise. The latest version of Java from java.sun.com will almost certainly work. At the time of writing this is version J2SE 6.0. It is essential that you install the **JDK** package and not just the JRE. We further recommend that you use the *Linux self-extracting file* version and not the *Linux RPM in self-extracting file* version. A typical installation would go like this:

```
cd /usr/local
sh jdk-6u14-linux-i586.bin
rm -f java; ln -s jdk1.6.0_14 java
```

This would install Java under /usr/local and replace any existing symbolic link called java with a new one.

You should then add /usr/local/java/bin to the start of your PATH to avoid other Java installations being picked up in preference and, if using a (t)csh shell, **rehash**. **Jemboss** installation examples in this book assume that the Java binaries are in /usr/local/java/bin unless otherwise specified.

1.7.2   Linux Debian distributions

These distributions include *Debian* itself and derivatives such as *Ubuntu*.

1.7.2.1 General prerequisites

Make sure you have the **X11** development files installed before you configure EMBOSS. You can use the following command to check:

```
dpkg --list x-dev
```

Install the **x-dev** package, if necessary, using the **dpkg -i** command or a graphical package manager.

1.7.2.2 PNG prerequisites

PNG support is optional; however, it is required if you intend using the **Jemboss** graphical interface. You need to make sure that the following libraries and development files have been installed from your distribution DVD/CDs:

- **zlib1g**
- **zlib1g-dev**
- **libpng12–0**
- **libpng12-dev**
- **libgd2-xpm**
- **libgd2-xpm-dev**

You can check to see they're there using the **dpkg --list** command as above. Use the **dpkg -i filename.deb** command to install them if necessary. Alternatively use a graphical package manager for installation.

1.7.2.3 PDF prerequisites

PDF support is optional. At the time of writing the libharu package is not provided with Debian distributions. You should therefore follow the 'Installing from source code' instructions (Section 1.10, 'PDF support: installing from source code').

1.7.2.4 Java

Everything that was said for Linux RPM distributions also holds for Debian distributions.

1.7.3 Mac OS X

These instructions refer to the 10.6 and 10.5 versions of Mac OS X although similar principles apply to 10.4 and 10.3.

Caution

Also note that the EMBOSS developers can only realistically support the package on a virgin Mac OS X installation.

However, there are two Mac OS X projects that provide pre-bundled EMBOSS distributions. They are the MacPorts project http://www.macports.org and the FINK project http://www.finkproject.org. If you choose to install EMBOSS from either of these sources then you will obviously not be following the standard UNIX methods for installation described earlier: in which case you should use the FINK or MacPorts documentation and support contacts (should anything go wrong). You may not require the information presented below, with the exception of the need for the **Xcode** tools. On the other hand, if you use (e.g.) the MacPorts **GD/PNG** libraries and are otherwise compiling the standard EMBOSS source code yourself, then by all means contact the EMBOSS developers for support if you have installation problems.

1.7.3.1   General prerequisites

It is essential to have installed the **Xcode** tools, **X11** (if provided) and any SDK packages from your distribution DVD. Not all of these may be installed by default so insert your DVD if necessary and do a *custom* installation and ensure their check boxes are ticked. To be sure of getting the latest version of (e.g.) **Xcode** then you will need to register as a developer at http://connect.apple.com.

1.7.3.2   PNG prerequisites

PNG support is optional. It is, however, essential if you intend installing the **Jemboss** graphical interface. You have the choice of downloading **gd** and **libpng** and their dependencies from the MacPorts or FINK projects, or compiling from source code. For the former see the instructions on the respective website, for the latter see below.

We recognise, from support queries, that there are a greater proportion of Mac OS X users who are unfamiliar with the UNIX command line compared to users of other UNIX flavours. Experienced hackers will have to forgive us for giving verbose installation instructions for this platform.

Installation of the PNG libraries and development files is done first, then the installation of the **gd** libraries and development files. Both are required for PNG support. So, first for PNG itself.

1. Download **libpng** from http://www.libpng.org/.
   We recommend you download the bz2 file e.g. `libpng-1.2.38.tar.bz2`: this contains a config script. It is useful to make a png directory (**mkdir png**) and put the file in there.
2. Uncompress the `libpng-1.2.38.tar.bz2` file:

```
bunzip2 libpng-1.2.38.tar.bz2
```

3. Untar the resulting tar file.

```
tar xf libpng-1.2.38.tar
```

4. This will create a `libpng-1.2.38` directory. Move into that directory.

```
cd libpng-1.2.38
```

5. Configure the software

```
./configure
```

6. Make the software

```
make
```

7. Install the software

```
make install
```

8. You can now delete the source code directory (libpng-1.2.38)

```
cd ..
rm -rf libpng-1.2.38
```

and, if you wish, the tar file.

```
rm libpng-1.2.38.tar
```

And now, for **gd**:

1. Download **gd** from http://www.libgd.org/.
   We recommend you download the gz file i.e. gd-2.0.35.tar.gz It is useful to make
   a **gd** directory (**mkdir gd**) and put the file in there.
2. Uncompress the gd-2.0.35.tar.gz file:

```
gunzip gd-2.0.35.tar.gz
```

3. Untar the resulting tar file.

```
tar xf gd-2.0.35.tar
```

4. This will create a gd-2.0.35 directory. Move into that directory.

```
cd gd-2.0.35
```

5. Configure the software using the standard GNU configure file in the gd-2.0.35
directory

```
./configure --without-freetype --without-fontconfig
```

Mac OS X does not ship with freetype or fontconfig support and EMBOSS does
not require them. If you want such support included then install their libraries
separately and adjust the configuration by removal of the --without statements.

6. Compile the software

```
make
```

7. Install the software

```
make install
```

8. You can now delete the source code directory (gd-2.0.35)

```
cd ..
rm -rf gd-2.0.35
```

and, if you wish, the tar file.

```
rm gd-2.0.35.tar
```

1.7.3.3   PDF prerequisites

PDF support is optional. At the time of writing the libharu package is not provided with
Mac OS X distributions. You should therefore follow the 'Installing from source code'
instructions (Section 1.10, 'PDF support: installing from source code').

1.7.3.4   Java

Java is optional; however, it is required if you intend using the **Jemboss** graphical user
interface. Both **Java** and the **Java SDK** are required and are provided on your Mac OS X
DVD. Note that they may not be installed using a standard system installation. Your Mac OS X
installation DVD may contain them so reinsert the DVD and do a *custom* installation as
necessary.

1.7.4   IRIX

1.7.4.1   General prerequisites

We recommend that you use the IRIX **cc** compiler as this will provide the greatest level of code optimization. The GNU **gcc** compiler is a suitable alternative. The EMBOSS configuration, like most `configure` scripts, will choose the **gcc** compiler by default if both compilers are available. To force use of **cc** you can type:

```
setenv CC cc [tcsh/csh]
CC=cc [sh/bash]
export CC
```

before configuring EMBOSS. There is a special configuration switch provided for the IRIX compiler in the EMBOSS configuration, namely:

```
--with-sgiabi=
```

This can have the values:

```
n32m3
n32m4
64m3
64m4
```

n32 refers to 32-bit processors and **64** to 64-bit processors. **m3** refers to the **Mips3** compiler and **m4** to the **Mips4** compiler. So, if you are using a 32-bit processor with the **Mips3** compiler then add the following switch to your configuration command:

```
./configure--with-sgiabi=n32m3 [ other optional switches ]
```

1.7.4.2   PNG prerequisites

PNG is optional. It is, however, required if you intend using the **Jemboss** graphical user interface.

The **libpng** and **gd** package from `freeware.sgi.com` are too out-of-date to work with EMBOSS. You should follow the 'Installing from source code' instructions (Section 1.9, 'PNG support: installing from source code').

1.7.4.3   PDF prerequisites

PDF support is optional. The **libharu** package is not provided with IRIX distributions. You should therefore follow the 'Installing from source code' instructions (Section 1.10, 'PDF support: installing from source code').

1.7.4.4   Java

Java is optional; however, it is required if you intend using the **Jemboss** graphical user interface.

Java for IRIX used to be available from the SGI Cool Software Index, however this has now been removed; you may have to search the internet to find a copy.

Caution

You must install the **JDK** and not just the **JRE**. Their latest version was 1.4.1_06 at the time of writing. This is acceptable for EMBOSS, but rather outdated. It will usually install into `/usr/java2`.

1.7.5   Tru64

1.7.5.1   General prerequisites

We recommend that you use the OSF1 **cc** compiler as this will provide the greatest level of code optimisation. The GNU **gcc** compiler is a suitable alternative. The EMBOSS configuration, like most configure scripts, will choose the **gcc** compiler by default if both compilers are available. To force use of **cc** you can type:

```
setenv CC cc [tcsh/csh]
CC=cc; export CC [sh/bash]
```

before configuring EMBOSS.

1.7.5.2   PNG prerequisites

PNG is optional. It is, however, required if you intend using the **Jemboss** graphical user interface. We recommend that you compile **zlib, libpng** and **gd** yourself by following the 'Installing from source code' instructions (Section 1.9, 'PNG support: installing from source code').

1.7.5.3   PDF prerequisites

PDF support is optional. The **libharu** package is not provided with Tru64 distributions. You should therefore follow the 'Installing from source code' instructions (Section 1.10, 'PDF support: installing from source code').

1.7.5.4   Java

Java is optional; however, it is required if you intend using the **Jemboss** graphical user interface. Java for Tru64 can be downloaded from http://www.hp.com/.

Caution

You must install the **JDK** and not just the **JRE**. The latest version was 1.4.2–08 at the time of writing. This is acceptable for EMBOSS, but rather outdated.

1.7.6   Solaris

These instructions refer to Solaris 10, although they should apply equally well to other versions of this operating system.

1.7.6.1   General prerequisites

We recommend that you use the Solaris **cc** compiler as this will provide the greatest level of code optimisation. The GNU **gcc** compiler is a suitable alternative. The EMBOSS configuration, like most configure scripts, will choose the **gcc** compiler by default if both compilers are available. To force use of **cc** you can type:

```
setenv CC cc [tcsh/csh]
CC=cc; export CC [sh/bash]
```

before configuring EMBOSS.

If you are using **gcc** then the version available on the companion CD is suitable. If using the **gcc** version from http://www.sunfreeware.com then you may need to rerun the header modification script fixincludes in that distribution.

/usr/ccs/bin must be in your path; otherwise the **ar** and **make** programs will not be found.

You should have the GNU **tar** program installed. The Solaris-supplied **tar** is too limited and may not extract the **Jemboss** files correctly. This can be downloaded from ftp://ftp.gnu.org/gnu/tar/:

```
./configure --prefix=/usr/local/
make install
```

Add /usr/local/bin to the start of your PATH and, if using a [t]csh shell, **rehash**.

1.7.6.2   PNG prerequisites

You should install PNG using the files supplied by http://www.sunfreeware.com/. The main packages are **gd** and **libpng**; however, there are many dependencies so you will probably find yourself having to install all these packages:

```
expat-2.0.1-sol10-sparc-local.gz
fontconfig-2.4.2-sol10-sparc-local.gz
freetype-2.3.9-sol10-sparc-local.gz
```

```
jpeg-6b-sol10-sparc-local.gz
libiconv-1.11-sol10-sparc-local.gz
xpm-3.4k-sol10-sparc-local.gz
libpng-1.2.38-sol10-sparc-local.gz
gd-2.0.35-sol10-sparc-local.gz
```

Use (e.g.)

```
gunzip expat-2.0.1-sol10-sparc-local.gz
pkgadd -d expat-2.0.1-sol10-sparc-local
```

The packages will be installed under the /usr/local directory tree.

Alternatively, to avoid all the dependencies, which EMBOSS does not require, you should consider compiling PNG support yourself using the 'Installing from source code" instructions (Section 1.9, 'PNG support: installing from source code'). If you do this then, to avoid conflict with sunfreeware files, install them in a directory tree other than /usr/local (e.g. use ./configure -prefix=/usr/local/png for the **PNG** packages and then configure EMBOSS using ./configure -with-pngdriver=/usr/local/png [+ *any other options]*).

1.7.6.3   PDF prerequisites

PDF support is optional. The **libharu** package is not provided with Solaris distributions. You should therefore follow the 'Installing from source code' instructions (Section 1.10, 'PDF support: installing from source code').

1.7.6.4   Java

Java is optional; however, it is required if you intend using the **Jemboss** graphical user interface.

Java for Solaris can be downloaded from http://java.sun.com/.

1.8   Troubleshooting EMBOSS installations

A commonly reported error is that the operating system does not recognise any EMBOSS commands, e.g. Command not found is reported when typing **embossversion**. The most common causes are that the compilation failed or that the PATH was incorrectly set. If you are unsure whether the compilation has failed then look for the executables in the bin directory appropriate for your configuration. For example, if you used - -**prefix=/usr/local/emboss** in your configuration then, after the **make install** look in the /usr/local/emboss/bin directory for the executables. If they are there then check your PATH as described above. If the executables are not there then study the output of the **make** command carefully, particularly the last few lines, as they should give you some idea why the compilation failed. It may be useful to recheck the prerequisite section of this chapter for your operating system.

Pre-EMBOSS 6.1.0 the most common compilation error was that, in mid-compilation, the compiler stopped and reported that -**lX11** could not be found. This was always because, whereas the operating system's **X11** server files had been installed, the **X11** development files hadn't. See the recommendations for **X11** in the prerequisites section of this chapter for your operating system if you get this error compiling older versions of EMBOSS. After installing the correct system files you should do a **make clean** and perform the **configure** command again.

The most recent distributions of EMBOSS test more extensively for the presence of **X11** files and should report their absence during the configuration stage. You therefore should not get the error mentioned in the paragraph above. If you do then please contact the EMBOSS development team emboss-bug@emboss.open-bio.org with details of your operating system and, as a file, the screen output from the configure command.

1.9   PNG support: installing from source code

These instructions should be followed if your system does not provide **zlib**, **libpng** and **gd** as optionally installable components. For example, all common Linux distributions do provide these components on the distribution DVD or CDs. Most other operating systems (e.g. IRIX and Solaris) provide **zlib** as standard but do not provide **libpng** or **gd**: they can usually be downloaded precompiled from freeware sites or installed from a companion CD distributed with the operating system. Unfortunately, such freeware has been known to be less than rigorously tested by the hardware companies, or it is too out-of-date to be usable by EMBOSS, and so you may find that you have no option but to compile the libraries yourself from source code.

Before compiling PNG support yourself, read any operating-system-specific notes for your machine in this chapter.

zlib, libpng and gd

It is imperative that and **zlib**, **libpng** and **gd** files are installed under the same directory tree. The exception is when some of the above components are in standard UNIX library directories and are provided along with the operating system. Standard system library directories usually have names like /lib or /usr/lib. As an example, on most UNIX operating systems, **zlib** is provided as standard: you therefore only have to install **libpng** and **gd** and those should be installed under their own common directory tree. Most UNIX software, when compiled from its source code, will install into the /usr/local directory tree. It's recommended you follow this tradition.

Caution

Note that /usr/local is not regarded as a 'standard system directory' as its name implies. If you already have some of the PNG components installed under the /usr/local directory tree then you should certainly install the rest of the components there as well.

PNG

The PNG components must be installed in the order **zlib**, **libpng** and finally **gd**.

zlib compilation. The procedure for **zlib** compilation is as follows.

1. Do a final check to see whether your system already has **zlib** installed.

(a) On Linux RPM based systems type:

```
rpm -q zlib
rpm -q zlib-devel  or just
rpm -qa "zlib*"
```

Both library and development packages should be shown. If not then install the relevant RPMs from your distribution DVD or CDs.

(b) On Linux Debian based systems type:

```
dpkg --list zlib1g
dpkg --list zlib1g-dev
```

Both should give a positive response. If not then install the relevant deb files from your distribution DVD or CDs.

(c) On other UNIX systems look in any system library directories (those named `lib` or `lib64` depending on your system) for a file called `libz.so`. A command that will usually find this file is:

```
find /usr -name libz.so -print
```

If the file is not found in a standard system directory then it's recommended you compile your own for use with EMBOSS.

2. Get the **zlib** source code

This can be downloaded from http://www.zlib.net/ . This assumes v1.2.3 is being used. Use whatever is the latest version of **zlib**. See the current versions section of this chapter.

3. In a temporary directory type:

```
gunzip zlib-1.2.3.tar.gz
tar xf zlib-1.2.3.tar
rm zlib-1.2.3.tar
cd zlib-1.2.3
```

4. Compile **zlib**:

```
./configure --shared --prefix=/usr/local
make
make install
```

This will install the library and development files under the `/usr/local` directory tree.

5. Delete the **zlib** source code:

```
cd ..
rm -rf zlib-1.2.3
```

1.9.1 LIBPNG compilation

1. Do a final check to see whether your system already has **libpng** installed.

   (a) On Linux RPM based systems type:

```
rpm -q libpng
rpm -q libpng-devel or just
rpm -qa "libpng*"
```

   Both library and development packages should be shown. If not then install the relevant RPMs from your distribution DVD or CDs.

   (b) On Linux Debian based systems type:

```
dpkg --list libpng12-0
dpkg --list libpng12-dev
```

   Both should give a positive response. If not then install the relevant deb files from your distribution DVD or CDs.

   (c) On other UNIX systems look in any system library directories (those named lib or lib64 depending on your system) for a file called libpng.so. A command that will usually find this file is:

```
find /usr -name libpng.so -print
```

   If the file is not found in a standard system directory then it's recommended you compile your own for use with EMBOSS.

2. Get the **libpng** source code.
   This can be downloaded from http://www.libpng.org/ . This example assumes that v1.2.38 is being used. Always use the latest version. See the current versions section of this chapter.

3. In a temporary directory type:

```
gunzip libpng-1.2.38.tar.gz
tar xf libpng-1.2.38.tar
rm libpng-1.2.38.tar
cd libpng-1.2.38
```

4. Compile **libpng**

```
./configure --prefix=/usr/local
make
make install
```

This will install the library and development files under the /usr/local directory tree.

5. Delete the LIBPNG source code:

```
cd ..
rm -rf libpng-1.2.38
```

1.9.2   gd **compilation**

1. Do a final check to see whether your system already has **gd** installed.

   (a) On Linux RPM based systems type:

```
rpm -q gd
rpm -q gd-devel  or just
rpm -qa "gd*"
```

Both library and development packages should be shown. If not then install the relevant RPMs from your distribution DVD or CDs.

   (b) On Linux Debian based systems type:

```
dpkg --list libgd2-xpm
dpkg --list libgd2-xpm-dev
```

Both should give a positive response. If not then install the relevant deb files from your distribution DVD or CDs.

   (c) On other UNIX systems look in any system library directories (those named lib or lib64 depending on your system) for a file called libgd.so. A command that will usually find this file is:

```
find /usr -name libgd.so -print
```

If the file is not found in a standard system directory then it's recommended you compile your own for use with EMBOSS.

2. Get the **gd** source code.

    This can be downloaded from http://www.libgd.org/ . This example assumes that v2.0.35 is being used. Always use the latest version. See the current versions section of this chapter. Versions of **gd** lower than 2.0.28 are completely unsuitable as they lack required GIF support.

3. In a temporary directory type:

```
gunzip gd-2.0.35.tar.gz
tar xf gd-2.0.35.tar
rm gd-2.0.35.tar
cd gd-2.0.35
```

4. Compile **gd**:

```
./configure --prefix=/usr/local --with-png=/usr/local --without-freetype --
without-fontconfig --without-jpeg --without-xpm
make
make install
```

This will install the library and development files under the /usr/local directory tree.

5. Delete the **gd** source code:

```
cd ..
rm -rf gd-2.0.35
```

> **Note**
>
> Note that **freetype, fontconfig,** jpeg and **xpm** functionality of the **gd** library has been disabled. These are not required for EMBOSS; only the PNG functionality is needed. If you have any of the other libraries installed then you can optionally amend the configuration line accordingly.

1.10   PDF support: installing from source code

These instructions should be followed if your system does not provide **libharu** as an optionally installable component. For example, Many common Linux RPM distributions do provide this component on the distribution DVD or CDs. Most other operating systems (e.g. IRIX and Solaris) don't. Sometimes it can be downloaded precompiled from freeware sites. Unfortunately, such freeware has been known to be less than rigorously tested by the hardware companies, or it is too out-of-date to be usable by EMBOSS, and so you may find that you have no option but to compile the libraries yourself from source code.

Before compiling PDF support yourself, read any operating-system-specific notes for your machine in this chapter.

PDF optional dependencies

The **libharu** PDF support package can provide optional PNG-handling support. If you require such support then install the **libpng** and **zlib** packages as described under 'PNG prerequisites' (Section 1.7, 'EMBOSS installation: platform-specific concerns') for your operating system. Note that, if you have compiled the **png** and **zlib** packages from source code then it makes sense to install the **libharu** package to the same prefix.

libharu compilation

The procedure for **libharu** compilation is as follows.

1. Do a final check to see whether your system already has **libharu** installed.
   (a) On Linux RPM based systems type:

```
rpm -q libharu
rpm -q libharu-devel or just
rpm -qa "libharu*"
```

Both library and development packages should be shown. If not then install the relevant RPMs from your distribution DVD or CDs.

   (b) On other UNIX systems look in any system library directories (those named `lib` or `lib64` depending on your system) for a file called `libhpdf.so`. A command that will usually find this file is:

```
find /usr -name libhpdf.so -print
```

If the file is not found in a standard system directory then it's recommended you compile your own for use with EMBOSS.

2. Get the **libharu** source code.
   This can be downloaded from http://www.libharu.org/ . This assumes v2.1.0 is being used. Use whatever is the latest version of **libharu**. See the current versions section of this chapter.

3. In a temporary directory type:

```
gunzip libharu-2.1.0.tar.gz
tar xf libharu-2.1.0.tar
rm libharu-2.1.0.tar
cd libharu-2.1.0
```

4. Compile **libharu**:

```
./configure --prefix=/usr/local
make
make install
```

This will install the library and development files under the /usr/local directory tree. If you do not require **libharu** to have PNG support then you can configure and install using:

```
./configure --prefix=/usr/local --without-zlib --without-png
make
make install
```

5. Delete the **libharu** source code:

```
cd ..
rm -rf libharu-2.1.0
```

1.11 Ncurses support: installing from source code

Ncurses support is only required if you intend installing one or both of the **mse** or **emnu** EMBASSY packages. These instructions should be followed if your system does not provide **Ncurses**, either by default or as optionally installable components. For example, all common Linux distributions do provide these components on the distribution DVD or CDs; Mac OS X also bundles **Ncurses** in its distribution. Most other operating systems (e.g. IRIX and Solaris) either do not, but **Ncurses** can usually be downloaded precompiled from freeware sites or installed from a companion CD distributed with the operating system. Unfortunately, such freeware has been known to be less than rigorously tested by the hardware companies and you may find that you have no option but to compile the libraries yourself from source code.

Before compiling **Ncurses** support yourself, read any operating-system-specific notes for your machine in this chapter.

1.11.1 Ncurses compilation

1. Do a final check to see whether your system already has **Ncurses** installed.

   (a) On Linux RPM based systems type:

```
rpm -q ncurses
rpm -q ncurses-devel or just
rpm -qa "ncurses*"
```

Both the library and development packages should be shown. If not then install the relevant RPMs from your distribution DVD or CDs.

(b) On Linux Debian based systems type:

```
dpkg -list libncursesw5
dpkg -list libncursesw5-dev
```

Both should give a positive response. If not then install the relevant deb files from your distribution DVD or CDs.

(c) On other UNIX systems look in any system library directories (those named lib or lib64 depending on your system) for files called libgd.so and ncurses.h. A command that will usually find these files would be:

```
find /usr -name libncurses.so -print
find /usr -name ncurses.h -print
```

If the files are not found in a standard system directory then it's recommended you compile your own for use with EMBOSS.

2. Get the **Ncurses** source code
   This can be downloaded from ftp.gnu.org/gnu/ncurses/. This example assumes that v5.7 is being used. Always use the latest version. See the current versions section of this chapter.

3. In a temporary directory type:

```
gunzip ncurses-5.7.tar.gz
tar xf ncurses-5.7.tar
rm ncurses-5.7.tar
cd ncurses-5.7
```

4. Compile **Ncurses**:

```
./configure --prefix=/usr/local
make
make install
```

This will install the library and development files under the /usr/local directory tree.

5. Delete the **Ncurses** source code:

```
cd ..
rm -rf ncurses-5.7
```

Building EMBASSY

2

2.1   Introduction to EMBASSY

The EMBASSY packages include applications with the same look and feel as
EMBOSS applications, but which are kept separate from EMBOSS. This is usually
because the packages are for specialised sequence analysis, or for non-sequence-based
analysis, or are licensed differently from EMBOSS (i.e. non-GPL), or done so at the
author's request.

The EMBASSY packages currently include:

- **PHYLIPNEW** (now **phylip** 3.68)
- **HMMERNEW**
- **EMNU**
- **ESIM4**
- **MEMENEW**
- **TOPO**
- **MSE**
- **VIENNA**

Also, various packages for protein structure and related areas:

- **DOMAINATRIX**
- **DOMALIGN**
- **DOMSEARCH**
- **SIGNATURE**
- **STRUCTURE**

2.2   Downloading the EMBASSY packages

The EMBASSY packages are available on the same FTP server and in the same directory as
the primary EMBOSS distribution. Follow the instructions in the EMBOSS chapter for
downloading instructions.

You cannot install the EMBASSY packages without first installing the main EMBOSS package, so do that before continuing. The EMBASSY packages can be installed from any temporary directory of your choice.

> **Note**
>
> **myemboss** is an EMBASSY package designed for use by EMBOSS software developers so you may not wish to install it if your site serves only biologists. **myembossdemo** contains usage examples of the **myemboss** package.

2.2.1   Unpacking the source code

You will have downloaded the EMBASSY source code to a suitable directory. Move to whichever directory you chose (e.g. **cd /usr/local/src/embassy**) and list the directory to make sure.

```
% ls
CBSTOOLS-1.0.0.tar.gz     HMMER-2.3.2.tar.gz        MYEMBOSSDEMO-6.1.0.tar.gz
DOMAINATRIX-0.1.0.tar.gz  IPRSCAN-4.3.1.tar.gz      PHYLIPNEW-3.68.tar.gz
DOMALIGN-0.1.0.tar.gz     MEMENEW-4.0.0.tar.gz      SIGNATURE-0.1.0.tar.gz
DOMSEARCH-0.1.0.tar.gz    MIRA-2.8.2.tar.gz         STRUCTURE-0.1.0.tar.gz
EMNU-1.05.tar.gz          MSE-1.0.0.tar.gz          TOPO-1.0.0.tar.gz
ESIM4-1.0.0.tar.gz        MYEMBOSS-6.1.0.tar.gz     VIENNA-1.7.2.tar.gz
```

These EMBASSY files are compressed binary files and require that your UNIX distribution has the **gunzip** program installed. Check that the command:

```
which gunzip
```

gives a positive response. If not then install the **gzip** package from whichever freeware site your UNIX distribution uses or, alternatively, compile the source code from the **gzip** homepage (http://www.gzip.org) which helpfully contains compilation instructions.

Taking **DOMAINATRIX** as an example, unpack it by typing:

```
gunzip DOMAINATRIX-0.1.0.tar.gz
```

This will create a file called DOMAINATRIX-0.1.0.tar. Such tar files are archive files containing the individual source code files. You must now extract the archive using the **tar** program:

```
tar xf DOMAINATRIX-0.1.0.tar
```

This will create a new directory, DOMAINATRIX-0.1.0: the exact name will depend on the version of **DOMAINATRIX** being unpacked.

> **Note**
>
> The **tar** program on most UNIX distributions will usually perform satisfactorily, however some will not. So, check the platform-specific notes (Section 2.8, 'EMBASSY installation: platform-specific concerns') section of this chapter to see whether you need to install the GNU version of the **tar** program instead.

Enter the directory and type `ls` to show the files. The directory listing should look something like this:

```
% cd DOMAINATRIX-0.1.0
% ls
aclocal.m4     config.sub      depcomp      ltmain.sh     missing
AUTHORS        configure       emboss_acd   m4            NEWS
ChangeLog      configure.in    INSTALL      Makefile.am   README
config.guess   COPYING         install-sh   Makefile.in   src
```

If it doesn't then you're in the wrong directory. This unpacking procedure applies to all the EMBASSY packages. You are now ready to build the packages.

2.3   Building the EMBASSY packages

This comprises the following steps:

1. Configuring
2. Compilation
3. Installation
4. Post-installation.

It goes without saying that your system must have a C compiler installed before attempting to build the EMBASSY packages. This will undoubtedly be the case as you will have built EMBOSS before reading this far. See Section 1.2, 'Building EMBOSS' if not.

2.4   Configuring the EMBASSY packages

Configuring EMBASSY packages is done using the configure script provided with each package (see the directory listing under Section 1.3, 'Unpacking the source code'). Each EMBASSY package must be configured before it is compiled.

There can be some pitfalls though, so look at the general prerequisites (Section 2.8, 'EMBASSY installation: platform-specific concerns') for your operating system before continuing.

2.4.1   A simple configuration

At its most simple all you should need to type (with suitable adjustment to the **--prefix** option) is:

```
./configure --prefix=/usr/local/emboss
```

In short, you must use exactly the same configuration options with which you configured the main EMBOSS package. There are two exceptions:

- You did not supply a --**prefix** option to the main EMBOSS package configuration or, equivalently, you used --**prefix=/usr/local** for the main EMBOSS package configuration. In both these cases EMBOSS will have been installed under the /usr/local directory tree. In these cases you must supply an extra option for the EMBASSY packages, namely -**enable-localforce**, i.e.

```
./configure --prefix=/usr/local -enable-localforce
```

- Some of the EMBASSY packages (**EMNU** and **MSE**) require an extra system library, namely the **Ncurses** library. If this library is installed in a non-standard place then you will need to configure with:

```
./configure --prefix=/usr/local/emboss --with-curses=DIR
```

where *DIR* specifies the location of the **Ncurses** library. For example, if the **Ncurses** library is in /opt/lib and the header files are in /opt/include then you would specify --**with-curses=/opt**.

The use of the --**prefix** option is vital: it tells the EMBASSY configuration where you installed the main EMBOSS package. If you cannot remember which options you supplied for the EMBOSS configuration then you can look in the config.status file at the top of the EMBOSS source code tree. The config.status file will have been created by the EMBOSS configuration. In the middle of that file you will see something similar to the following lines:

```
% more config.status
..
configured by ./configure, generated by GNU Autoconf 2.63,
with options \"'--prefix=/usr/local/emboss'\"
..
```

2.4.1.1   CBSTOOLS

You need only specify the compilation options you used for the main EMBOSS package. The **CBS** packages chlorop, lipop, netnglyc, netoglyc, netphos, prop, signalp, tmhmm and yinoyang are required for the corresponding EMBOSS wrappers provided by **CBSTOOLS**.

2.4.1.2   DOMAINATRIX

You need only specify the compilation options you used for the main EMBOSS package.

2.4.1.3 DOMALIGN

You need only specify the compilation options you used for the main EMBOSS package.

2.4.1.4 DOMSEARCH

You need only specify the compilation options you used for the main EMBOSS package.

2.4.1.5 EMNU

You need to specify both the configuration options used for the main EMBOSS package and may also require the `--with-curses=DIR` option. See the platform-specific notes (Section 2.8, 'EMBASSY installation: platform-specific Concerns') for **Ncurses**.

EMNU was written as a simple text-based interface before the advent of GUIs for EMBOSS. Various interfaces are now available (see the *EMBOSS User's Guide*).

2.4.1.6 Esim4

You need only specify the compilation options you used for the main EMBOSS package. The **sim4** package is required.

2.4.1.7 Hmmer

You need only specify the compilation options you used for the main EMBOSS package. The **hmmer** package is required.

2.4.1.8 Iprscan

You need only specify the compilation options you used for the main EMBOSS package. The **InterProScan** package is required. The **CBS** packages **signalp** and **tmhmm** can optionally be installed.

2.4.1.9 Meme

You need only specify the compilation options you used for the main EMBOSS package. The **meme** package is required.

2.4.1.10 Mira

You need only specify the compilation options you used for the main EMBOSS package. The **mira** package is required.

2.4.1.11 Mse

You need to specify both the configuration options used for the main EMBOSS package and you may also need the `--with-curses=DIR` option. See the platform-specific notes (Section 2.8, 'EMBASSY installation: platform-specific concerns') for **Ncurses**.

2.4.1.12 Myemboss

You need only specify the compilation options you used for the main EMBOSS package. Note that this package is meant to be installed only by EMBOSS developers. Each must have their own copy. It is not the job of the system administrator to install this package.

2.4.1.13   Myembossdemo

You need only specify the compilation options you used for the main EMBOSS package.

2.4.1.14   Phylip

You need only specify the compilation options you used for the main EMBOSS package.

2.4.1.15   Signature

You need only specify the compilation options you used for the main EMBOSS package.

2.4.1.16   Structure

You need only specify the compilation options you used for the main EMBOSS package.

2.4.1.17   Topo

You need only specify the compilation options you used for the main EMBOSS package.

2.4.1.18   Vienna

You need only specify the compilation options you used for the main EMBOSS package.

2.5   Compiling the EMBASSY packages

After you have configured an EMBASSY package then the compilation stage consists of typing only one command:

```
make
```

This should be done in the same directory from which you invoked the **configure** command. If you have a multiprocessor machine then you can speed up the compilation by specifying how many processors to use, e.g.

```
make -j 4
```

will specify the use of four processors (or two processors with hyperthreading enabled).
   After compilation has completed successfully you will see something similar to the following at the bottom of your screen.

```
..
make[2]: Leaving directory '/usr/local/src/embassy/DOMAINATRIX-0.1.0/src'
make[1]: Leaving directory '/usr/local/src/embassy/DOMAINATRIX-0.1.0/src'
Making all in emboss_acd
make[1]: Entering directory '/usr/local/src/embassy/DOMAINATRIX-0.1.0/emboss_acd'
make[1]: Nothing to be done for 'all'.
make[1]: Leaving directory '/usr/local/src/embassy/DOMAINATRIX-0.1.0/emboss_acd'
```

```
make[1]: Entering directory '/usr/local/src/embassy/DOMAINATRIX-0.1.0'
make[1]: Nothing to be done for 'all-am'.
make[1]: Leaving directory '/usr/local/src/embassy/DOMAINATRIX-0.1.0'
%
```

2.5.1 Reporting compilation errors

If the compilation failed then recheck the platform-specific notes (Section 2.8, 'EMBASSY installation: platform-specific concerns') for your operating system to make sure you have all the prerequisites installed; then reconfigure and try again assuming you spotted a mistake. If all else fails then email emboss-bug@emboss.open-bio.org for help including:

- The output from the **configure** command
- The screen output from the **make** command (not just the last few lines)
- The config.status file
- The config.log file

You can capture the output from (e.g.) the **make** command using:

```
make >&! make.out [csh shells] or
make > make.out 2>&1 [sh shells]
```

This can be done for the **configure** command too.

2.6 Installing the EMBASSY packages

After having successfully configured and compiled an EMBASSY package then all the libraries and applications need installing to the directory tree you specified using the - -**prefix** option to the **configure** command. This is done using one command:

```
make install
```

In fact, if the compilation had not already been done by typing **make** alone, then the above command would perform both the compilation and installation. We recommend, however, that you separate the two phases as we've described them; it helps in detecting where any failures may have happened.

Similarly to the compilation phase, many hundreds of lines will scroll up the screen. It is very rare for there to be errors in the installation phase as long as you checked that you had write permission in the directory tree you specified using - -**prefix** in the configuration. If all has gone well then the last few lines of the installation output will look something like this.

```
make[2]: Leaving directory '/usr/local/src/embassy/DOMAINATRIX-0.1.0/emboss_acd'
make[1]: Leaving directory '/usr/local/src/embassy/DOMAINATRIX-0.1.0/emboss_acd'
make[1]: Entering directory '/usr/local/src/embassy/DOMAINATRIX-0.1.0'
```

```
make[2]: Entering directory '/usr/local/src/embassy/DOMAINATRIX-0.1.0'
make[2]: Nothing to be done for 'install-exec-am'.
make[2]: Nothing to be done for 'install-data-am'.
make[2]: Leaving directory '/usr/local/src/embassy/DOMAINATRIX-0.1.0'
make[1]: Leaving directory '/usr/local/src/embassy/DOMAINATRIX-0.1.0'
%
```

2.7   Post-installation of the EMBASSY packages

No post-installation should be necessary as you will have followed the post-installation instructions when installing the main EMBOSS package. Of course, if you are using a csh shell variant then you'll have to type **rehash** to access the installed programs, or log out and back in again.

2.7.1   Testing an EMBASSY installation

An easy way to check that the EMBASSY packages have been installed is to use the EMBOSS application **wossname**.

```
% wossname

Finds programs by keywords in their one-line documentation
Keyword to search for, or blank to list all programs:


..
PROTEIN 3D STRUCTURE
domainreso   Remove low resolution domains from a DCF file
..

UTILS DATABASE CREATION

..
domainnr     Removes redundant domains from a DCF file
domainseqs   Adds sequence records to a DCF file
domainsse    Add secondary structure records to a DCF file
..
scopparse    Generate DCF file from raw SCOP files
ssematch     Search a DCF file for secondary structure matches
..
```

The abbreviated output above shows that the **DOMAINATRIX** programs have indeed been installed.

2.7.2   Deleting an EMBASSY package

If you've kept the source code tree from which you'd done the **make install** then, to delete a specific EMBASSY package, go to the top level of its source code tree and type:

```
make uninstall
```

If you didn't keep the source code tree then the preferred method of uninstallation is to reinstall the EMBASSY package on top of itself and then uninstall as above.

2.7.3 Keeping EMBASSY up to date

Any update to an existing EMBASSY package will have a new version number. It will be found at the top level of our FTP server.

```
ftp://emboss.open-bio.org/pub/EMBOSS/
```

You should delete the existing EMBASSY package and reinstall the new one.

2.7.4 Reinstallation of EMBASSY and EMBOSS

If you install a new version of EMBOSS then you should also download and reinstall all the EMBASSY packages. The EMBASSY packages on our FTP server are tailored to match the current EMBOSS version.

> **Note**
>
> There will not usually be added functionality in such an EMBASSY package therefore its version number may not change.

2.8 EMBASSY installation: platform-specific concerns

2.8.1 Linux RPM distributions

These distributions include Fedora Core, SuSE, Mandriva and RedHat.

2.8.1.1 General prerequisites

See the platform-specific notes (Section 1.7, 'EMBOSS installation: platform-specific concerns') for installing the main EMBOSS package.

2.8.1.2 Ncurses prerequisites

You need to make sure that the following libraries and development files have been installed from your distribution DVD/CDs:

```
ncurses
ncurses-devel
```

You can check whether they're there using the **rpm -q** *PackageName* command as above. Use the **rpm -i** *PackageName*.**rpm** command to install them if necessary (alternatively use **yum install** *PackageName* or a graphical package manager if available).

2.8.2   Linux Debian distributions

These distributions include Debian itself and derivatives such as Ubuntu.

2.8.2.1   General prerequisites

See the platform-specific notes (Section 1.7, 'EMBOSS installation: platform-specific concerns') for installing the main EMBOSS package.

2.8.2.2   Ncurses prerequisites

You need to make sure that the following libraries and development files have been installed from your distribution DVD/CDs:

```
libncursesw5
libncursesw5-dev
```

You can check whether they're there using the **dpkg --list** command. Use the **dpkg -i filename.deb** command to install them if necessary (alternatively use a graphical package manager).

2.8.3   Mac OS X

These instructions refer to the 10.6 and 10.5 versions of Mac OS X although similar principles apply to 10.4 and 10.3.

Caution

The EMBOSS developers can only realistically support the package on a virgin Mac OS X installation.

2.8.3.1   General prerequisites

See the platform-specific notes (Section 1.7, 'EMBOSS installation: platform-specific concerns') for installing the main EMBOSS package.

2.8.3.2   Ncurses prerequisites

This package comes with Mac OS X so no installation is necessary.

2.8.4   IRIX

2.8.4.1   General prerequisites

See the platform-specific notes (Section 1.7, 'EMBOSS installation: platform-specific concerns') for installing the main EMBOSS package.

2.8.4.2 Ncurses prerequisites

The EMBOSS configuration has been tailored to accept the **Ncurses** package available from http://www.freeware.sgi.com (as long as the **cc** compiler is being used and the configuration command contains a suitable **-with-sgiabi** switch). This is, at the time of writing:

```
ncurses-5.3
```

It can be installed using **inst** and will be loaded into the /usr/freeware directory tree and so would be selected from the EMBASSY configuration using:

```
-with-curses=/usr/freeware
```

As you can see, it is rather out of date. It may therefore be advisable to compile it yourself by following the instructions (Section 1.11, '**Ncurses** support: installing from source code'). If you are using the **gcc** compiler then you should most certainly compile **Ncurses** support yourself.

2.8.5 Tru64

2.8.5.1 General prerequisites

See the platform-specific notes (Section 1.7, 'EMBOSS installation: platform-specific concerns') for installing the main EMBOSS package.

2.8.5.2 Ncurses prerequisites

We recommend that you compile **Ncurses** yourself by following the instructions (Section 1.11, '**Ncurses** support: installing from source code'). You can then specify its location in the EMBASSY configuration using (e.g.):

```
--with-curses=/usr/local
```

2.8.6 Solaris

These instructions refer to Solaris 10, although they should apply equally well to other versions of this operating system.

2.8.6.1 General prerequisites

See the platform-specific notes (Section 1.7, 'EMBOSS installation: platform-specific concerns') for installing the main EMBOSS package.

2.8.6.2   Ncurses prerequisites

Some **Ncurses** packages are provided on companion CDs and on the sunfreeware site. Both of these are problematic: it's therefore strongly recommend that you install **Ncurses** from source code using the instructions (Section 1.11, '**Ncurses** support: installing from source code').

Building Jemboss

3

3.1   Introduction to Jemboss

Jemboss is a graphical user interface (GUI) for the EMBOSS package. It can be configured to be a client-server GUI or run as a standalone GUI on a workstation or home PC. It is written in Java.

3.1.1   Jemboss as a client-server

In this mode of operation, **Jemboss** and EMBOSS are installed on a server machine. All the EMBOSS applications will run on this server. The server must be a UNIX machine. Other computers on the network can request a copy of the GUI from the server. The GUI is self-contained: it comprises not only the user interface but also all the communication software for contacting the server and running EMBOSS applications. The client machines can be running any operating system for which there is Java support.

The server can be configured to allow anyone to run the EMBOSS applications (non-authenticating) or require that a GUI user must have an account on the server and that a username and password is typed at the start of the GUI session (authenticating).

The **Jemboss** server can also be configured to communicate with clients either via standard ASCII or to use automatic encryption (SSL) for security.

3.1.2   Jemboss as a standalone GUI

This mode of operation is used on self-contained EMBOSS installations. These will typically be a home PC, a lab PC or a laptop. The **Jemboss** GUI and EMBOSS itself are installed on the same machine.

3.2   Installing as a standalone GUI

This method of setting up **Jemboss** is designed for local copies of EMBOSS on either a workstation or laptop. From EMBOSS v6.1.0 onwards an installation of the main EMBOSS package (Chapter 1, *Building EMBOSS*) will automatically configure a standalone version of **Jemboss**.

3.2.1   Prerequisites

- A UNIX workstation or laptop
- SUN **Java JDK** or **JRE** from http://java.sun.com, or OpenJDK.
- A fresh unpacking of the EMBOSS source code, i.e. remove all existing versions.
- PNG graphics (see platform-specific notes (Section 1.7, 'EMBOSS installation: platform-specific concerns') in the EMBOSS installation instructions.)

3.2.2   Current software versions

At the time of writing the following software versions are current and will work with **Jemboss**. For the lifetime of the project we will, as far as possible, make **Jemboss** compatible with the latest stable Java releases.

- Java 6 JDK
- OpenJDK 6

3.2.3   Compiling EMBOSS for use with the standalone Jemboss

All you need to do is to follow the instructions for compiling the main EMBOSS package (Chapter 1, *Building EMBOSS*). However, you must have ended up installing it using a **make install** command. As well as installing the EMBOSS binaries to the bin installation subdirectory this will also create a script for running the standalone **Jemboss**. The script is called **runJemboss.csh** and can be run, by name, just like any other EMBOSS application or UNIX system command.

3.2.4   Post-installation of the standalone GUI

As long as you have followed the instructions for the post-installation of the main EMBOSS package (Section 1.6, 'Post-installation of EMBOSS'), with particular reference to correctly setting the PATH, then no further post-installation should be required.

3.3   Installing as an authenticating client-server

3.3.1   Prerequisites

- A UNIX server
- Accounts on the server for all users who will use the GUI from a remote client
- SUN **Java JDK** (the **JRE** is not enough) from http://java.sun.com or OpenJDK
- Apache **Tomcat** from http://tomcat.apache.org
- Apache **Axis** from http://ws.apache.org/axis/
- A non-privileged user account on the server, which will run **Tomcat/Jemboss**
- A fresh unpacking of the EMBOSS source code
- A working HTTP server
- An open port for use by the **Jemboss** server (default 8443)

- PNG graphics: see platform-specific notes (Section 1.7, 'EMBOSS installation: platform-specific concerns') in the EMBOSS installation instructions
- Completely remove any **Tomcat** installation and EMBOSS source and installation directories, likely to conflict, each time you do an installation
- For batch use see elsewhere (Section 3.9, 'Setting up **Jemboss** to use batch queuing software').

Important

It is important to completely remove any **Tomcat** installation and EMBOSS source and installation directories, likely to conflict, each time you do an installation. This is because the installation script modifies the content of these directories and deletion/re-extraction is necessary to properly reinitialise them. Attempting to retry an installation over the top of a failed one is doomed to cause another failure. Also check that you haven't got your **Tomcat** server running before starting the installation process.

3.3.2 Current software versions

At the time of writing the following software versions are current and will work with **Jemboss**.

- **Tomcat** 5.5.27
- **Axis** 1.4
- Java 6 **JDK** or **OpenJDK** 6. The minimum requirement is v1.4.1. However, for anything less than Java 5, you need to install the **Tomcat** compatibility binaries on top of the base **Tomcat** binaries

3.3.3 Client-server installation

3.3.3.1 Setting up EMBASSY packages

If you wish to use the EMBASSY packages with the **Jemboss** GUI then you must download and unpack them first. This must be done in a top-level sub-directory called embassy in the EMBOSS source tree. For example, to have ESIM4 and TOPO installed along with EMBOSS:

```
gunzip EMBOSS-6.1.0.tar.gz
   tar xf EMBOSS-6.1.0.tar
   cd EMBOSS-6.1.0
   mkdir embassy (the directory must have this name and location)
   cd embassy
   ftp emboss.open-bio.org (retrieve for example ESIM4 and TOPO sources and unpack)
```

You should not configure the EMBASSY packages as this will be done as part of the **Jemboss** installation.

3.3.3.2 Installing Tomcat and Axis

This should be done as the unprivileged user you'll have created for running the server. This username must not have a UNIX gid (group ID) of '0' otherwise the server will not work: this is a security measure. You should download the binary versions of these packages. Assuming the username you're going to use is called `jemboss`:

```
cd /fu/bar (this assumes user jemboss has permission to write here)
gunzip apache-tomcat-5.5.27.tar.gz
tar xf apache-tomcat-5.5.27.tar
gunzip axis-bin-1_4.tar.gz
tar xf axis-bin-1_4.tar
```

This will have created the directories `apache-tomcat-5.5.27` and `axis-1_4`.

3.3.3.2.1 Creating installation and results directories

You need to decide in which directory the final server (not the source code you're creating it from) should live and make sure, before running the installation script, that this directory exists and that the non-privileged installation user has write access to it. Following UNIX tradition an obvious directory to create for this is `/usr/local/emboss`, but you can choose anywhere in your filing system.

There is one caveat: it is recommended that the server be installed to directly attached disk rather than disk attached via (e.g.) NFS. Network-accessed disk can, in some cases, have a very detrimental effect on **Tomcat** server performance. So:

```
mkdir /usr/local/emboss
chown jemboss /usr/local/emboss
```

You also need to decide where the user results will be kept on the server. The server will automatically create sub-directories under this location for each user who runs an EMBOSS application. How much space is required will, of course, depend on how many users you have and how much use they'll make of the server. You will have to devise your own strategy for managing this disk space, e.g. whether you leave user results there permanently or delete them after a certain period. The directory must be world writeable and it should have the sticky bit set, e.g.:

```
mkdir /fu/bar/results
chmod 1777 /fu/bar/results
```

3.3.3.2.2 Running the installation script

We recommend installing **Jemboss** using the installation script supplied as part of the `EMBOSS-x.y.z.tar.gz` bundle. It lives in the sub-directory `jemboss/utils` and is called `install-jemboss-server.sh`.

This script should be run as an unprivileged user. The **Tomcat** server associated with **Jemboss** will end up being run under this username.

In most cases the installation script will produce a default response. If this is the value required then you can just press <RETURN>. The examples below always show a fully typed response for clarity.

Here is a typical run for setting up an authenticating server:

```
gunzip EMBOSS-6.1.0.tar.gz
tar xf EMBOSS-6.1.0.tar
cd EMBOSS-6.1.0/jemboss/utils
./install-jemboss-server.sh
```

```
- - - - - - - - - - - - - - - - - - - - - - - - - - - - - - - - - - - - - - - - - - - - - - - - - - - - - - - - -
EMBOSS and Jemboss Server installation script
- - - - - - - - - - - - - - - - - - - - - - - - - - - - - - - - - - - - - - - - - - - - - - - - - - - - - - - - -

Note: any default values are given in square brackets [].

This script installs EMBOSS as well as Jemboss.
Jemboss is deployed as a Java web application in your tomcat server.
A script is prepared to run Jemboss client that by default uses the
above Jemboss web application.

For detailed information on installing jemboss see:
http://emboss.open-bio.org/Jemboss/install/setup.html

*** This script needs to be run with permissions to be able
*** to install EMBOSS in the required directories. This may
*** be best done as root or as a tomcat user.

Before running this script you should download the latest:

(1) EMBOSS release (contains Jemboss) ftp://emboss.open-bio.org/pub/EMBOSS/
(2) Tomcat release http://tomcat.apache.org/
(3) Apache AXIS (SOAP) release 1.4 http://ws.apache.org/axis/

Have the above been downloaded (y/n)?
y
```

These packages are part of the prerequisites given above so ought to have been downloaded and installed at this stage, so y has been typed. Typing n would abort the installation.

```
Select the platform that your Jemboss server will be
run on from 1-8 [1]:
(1) linux
(2) aix
(3) irix
(4) hp-ux
(5) solaris
(6) macosX
(7) OSF
(8) FreeBSD
1
```

This example installation assumes the Linux operating system is being used.

```
The IP address is needed by Jemboss to access
the Tomcat web server.
Enter IP of server machine [localhost]:
192.168.8.11
```

Whereas the default value of localhost will work in most cases it is always preferable to use the full IP address of the server; 192.168.8.11 in this example. Using the full IP address always works, as long as it's the right address.

```
Enter if you want the Jemboss server to use data
encryption (https/SSL) (y,n) [y]?
y
```

It is difficult to see why you would not want to use data encryption. The overhead in doing so is very slight. You should certainly use data encryption with any authenticating server.

```
Enter port number [8443]
8443
```

8443 is the default port through which the two-way communication occurs between the client and the server. You may change this here but you must make sure that whatever port you choose is both free and not disallowed by any firewall you have in operation.

```
Enter java (1.4 or above) location [/usr]:
/usr/local/java
```

This is the location of the top-level directory of your SUN or OpenJDK **Java JDK** installation.

```
Enter EMBOSS download directory
[/ml/EMBOSS-6.1.0]:
/ml/EMBOSS-6.1.0
```

This is the top-level directory of the EMBOSS source code after extraction from the compressed tar file. The installation script works out the default value as being two directories above the location from where the installation script was invoked. In this example the script was invoked from /ml/EMBOSS-6.1.0/jemboss/utils/ and the suggested value is correct.

```
Enter where EMBOSS should be installed [/usr/local/emboss]:
/usr/local/emboss
```

This is where the compiled server will be installed, described under 'Creating installation and results directories' (Section 3.3.3.2.1). Following UNIX convention the default location is under /usr/local.

```
Enter URL for emboss documentation for application
[http://emboss.open-bio.org/]:
http://emboss.open-bio.org/
```

Jemboss will look for EMBOSS application documentation over the network. The default URL is the main documentation site. It should only be changed if you wish to use a mirror of the main site.

```
Do you want Jemboss to use unix authorisation (y/n) [y]?
y
```

This is an authenticating server installation so the answer is y. If you answer n then anyone will be able to connect to the server; usernames and passwords are not required.

```
Provide the UID of the account (non-privileged) to run Tomcat,
it has to be greater than 100 [501]:
501
```

You are, or should be, running this installation script under the username that will be running the **Jemboss/Tomcat** server. That being the case, the installation script will try to determine the UID associated with the current username and set the default response appropriately. In the above example, the default is correct i.e. the installation script has found the appropriate UID.

```
Unix Authentication Method, see:
http://emboss.open-bio.org/Jemboss/install/authentication.html
(1) shadow     (3) PAM         (5) HP-UX shadow
(2) no shadow (4) AIX shadow (6) Re-entrant shadow
(7) Re-entrant no shadow
Type of unix password method being used
(select 1, 2, 3, 4, 5, 6 or 7) [3]
3
```

This is, perhaps, the most technical question in the installation. Suffice it to say for now that the installation script will usually set the correct default. It does this based on what you typed in response to the operating system question earlier. As a heuristic, if your operating system has PAM then choose it.

```
Define the directory you want to store the results in
[/tmp/SOAP/emboss]
/fu/bar/results
```

This is the results directory described under 'Creating installation and results directories'. You almost certainly do not want to accept the default location as the /tmp filesystem is usually limited and is often wiped clean after a system reboot.

```
Enter Tomcat root directory (e.g. /usr/local/tomcat)
/fu/bar/apache-tomcat-5.5.27
```

This is the location of **Tomcat** as described under 'Installing **Tomcat** and **Axis**' (Section 3.3.3.2).

```
Enter Apache AXIS (SOAP) root directory (e.g. /usr/local/axis)
/fu/bar/axis-1_4
```

This is the location of **Axis** as described under 'Installing **Tomcat** and **Axis**' (Section 3.3.3.2).

```
-------------------------------------------------------------
The libraries for EMBOSS (libpng and gd) appear to be in /usr,
if these are the correct libraries then there should be no need
to add any configuration options.
Enter any other EMBOSS configuration options (e.g. --with-pngdriver=pathname)
or press return to leave blank:
```

At this point the installation script has tested to see whether it can find the **png/gd/zlib** headers on your system. PNG graphics are required for **Jemboss** to display and store any graphical output from EMBOSS. The script will look for the headers under /usr, /usr/ local, /opt/freeware and /usr/freeware. If, however, the script cannot find your PNG installation then you should enter it here.

> **Note**
>
> If your PNG installation is in **/fu/bar/include** and **/fu/bar/lib** then you would use **--with-pngdriver=/fu/bar** as the configuration option.

This prompt is also an opportunity to add any special compiler or linker flags peculiar to your operating system. On many operating systems you will not need to do this. This is the case in this example so <RETURN> was pressed.

At this point the script will configure, compile and install EMBOSS and **Jemboss**. You will see hundreds of lines scrolling up your terminal screen as this happens. Depending on the speed of your machine this may take 5–60 minutes. Eventually you will receive the following prompt.

```
-------------------------------------------------------------
EMBASSY packages can optionally be installed along with
the EMBOSS applications, see:
http://emboss.open-bio.org/rel/dev/apps/
where 'release' corresponds to the EMBOSS version e.g. "R3", "R4", "R5" etc.
-------------------------------------------------------------
Install EMBASSY packages (y,n) [y]?
n
```

The 'Installation of EMBASSY packages' subsection above described the pre-installation steps you need to do if you wish to install EMBASSY applications under **Jemboss**. If you require EMBASSY package installation then type 'y' at this prompt. You will then be given the option to install all such prepared packages or to choose them individually; the latter option is useful for developers as it allows non-EMBASSY directories to be skipped (e.g. CVS directories).

This simple installation example chooses not to install any EMBASSY packages.

After the selection of any EMBASSY packages the installation script will then configure, compile and install them either en masse or individually. You will then return to the general installation script prompts.

```
------------------------- ClustalW -------------------------
To use emma (EMBOSS interface to ClustalW) Jemboss needs to
know the path to the clustalw binary.
Enter the path to clustalw or press return to set
this later in jemboss.properties
/usr/local/bin
```

The **clustalw** binary is required by the EMBOSS application **emma**. You can either enter the location as above or add it to the `jemboss.properties` file at a later stage. The latter file is described elsewhere in this chapter.

```
------------------------- Primer3 -------------------------
To use eprimer3 (EMBOSS interface primer3 from the Whitehead
Institute) Jemboss needs to know the path to the primer3_core
binary.
Enter the path to primer3_core or press return to set
this later in jemboss.properties
/usr/local/bin
```

The **primer3_core** binary is required by the EMBOSS application **eprimer3**. You can either enter the location as above or add it to the `jemboss.properties` file at a later stage. The latter file is described elsewhere in this chapter.

After this prompt the installation scripts will create `wossname.jar`, `jemboss.properties` and manifest files. Many lines will scroll up your terminal screen.

If you have selected https encryption you will now be asked for a password.

```
-------------------------------------------------------------
Client and server certificates need to be generated for the
secure (https) connection. These are then imported into
keystores. The keystores act as databases for the
certificates.
for details see:
http://emboss.open-bio.org/Jemboss/install/ssl.html
-------------------------------------------------------------
Provide a password (must be at least 6 characters):
helloworld
```

This password will end up in the `server.xml` file of the **Tomcat** server and, whereas it will not be visible to non-privileged users, it will be visible to other system administrators on your server.

```
Provide the validity period for these certificates, i.e. the
number of days before they expire and new ones need to be made [90]:
90
```

This question is self-explanatory. Elsewhere in this chapter it is described how to create new certificates once the initial ones expire.

At this point the certificates will be created and more lines will scroll up your terminal screen.

```
Tomcat XML deployment descriptors have been created for the Jemboss Server.
Would you like an automatic deployment of these to be tried (y/n) [y]?
y
```

This is a sensitive point in the installation. You should certainly try to deploy the XML at this stage, hence the prompt has been answered with y. The installation script will now print some XML to the screen at this point. You should follow the instructions on screen to copy it into your **Tomcat** configuration directory. There is usually a rather obvious place to paste the XML, namely before the line which says:

```
Define a SSL HTTP/1.1 Connector on port 8443
```

You will see that the definition of that connector has been commented out so there is no danger of the definitions clashing.

Be aware that the `server.xml` file mentioned below also defines an http connector on port 8080. If that number is already in use on your system then you should change it.

```
1) COPY & PASTE THE FOLLOWING INTO
   /fu/bar/apache-tomcat-5.5.27/conf/server.xml

   Connector port="8443" minProcessors="5" maxProcessors="75"
           enableLookups="false"
           acceptCount="10" debug="0" scheme="https" secure="true"
           useURIValidationHack="false"

keystoreFile="/usr/local/emboss/share/EMBOSS/jemboss/resources/server.keystore"
keystorePass="helloworld"
           clientAuth="false" sslProtocol="TLS"/>

To continue you must have changed the above file(s)!
Have the above files been edited (y/n)?
y
```

Before typing y it is as well to check that no **Jemboss** server is currently running from a previous installation otherwise deployment will fail. Use the UNIX command **ps** in a separate terminal to check for any likely server processes.

After typing y the following text will appear on screen.

```
Please wait, starting tomcat.......
Using CATALINA_BASE:   /fu/bar/apache-tomcat-5.5.27
Using CATALINA_HOME:   /fu/bar/apache-tomcat-5.5.27
Using CATALINA_TMPDIR: /fu/bar/apache-tomcat-5.5.27/temp
Using JRE_HOME:        /usr/local/java
```

At this point the **Tomcat** server will have been started but there may be a 30-second pause before the deployment is attempted. This gives the server time to finish its initialisation.

When the deployment happens the following text should appear. The WARNING on the second line should be ignored.

```
Jan 8, 2007 4:03:40 PM org.apache.axis.utils.JavaUtils isAttachmentSupported
WARNING: Unable to find required classes (javax.activation.DataHandler and
javax.mail.internet.MimeMultipart). Attachment support is disabled.
Processing file JembossServer.wsdd
Done processing

-----------------------------------------------------------------------
-----------------------------------------------------------------------

Tomcat should be running and the Jemboss web services deployed!
(see https://192.168.8.11:8443/axis/)

It is *very* important to now:
1. As root:
    chown root /usr/local/emboss/bin/jembossctl
    chmod u+s /usr/local/emboss/bin/jembossctl
2. Ensure that tomcat is running as the non-privileged user,
    with the same UID (i.e. 501) that was given to this script
    (and NOT as root!).
3. Use the tomstop & tomstart scripts in this directory
    to stop & start tomcat.

Try running Jemboss with the script:
    /usr/local/emboss/share/EMBOSS/jemboss/runJemboss.sh

To create a web launch page see:
http://emboss.open-bio.org/Jemboss/install/deploy.html
```

If the deployment fails then it will probably do so with a Java stack trace, in which case look at the troubleshooting section (Section 3.11, 'Troubleshooting **Jemboss** installation'). The server relies on a setuid root program to be able to masquerade as a remote user. These permission changes can only be done by root:

```
chown root /usr/local/emboss/bin/jembossctl
chmod u+s /usr/local/emboss/bin/jembossctl
```

At this point the **Tomcat** server will be running and can be tested using a script created during the installation. As the unprivileged user try out the **Jemboss** interface by typing:

```
/usr/local/emboss/share/EMBOSS/jemboss/runJemboss.sh
```

Navigate, within the interface, to the embossversion application and click on GO. If all is well then a window will pop up showing the EMBOSS version being used.

3.3.3.3   Adding EMBASSY packages to the server (post-installation)

It is usually more convenient to add EMBASSY packages to the **Jemboss** server by selecting them from the install-jemboss-server.sh script as previously described. It is,

75

however, possible to add them afterwards using the instructions given in the EMBASSY installation chapter (Chapter 2, *Building EMBASSY*). The **Jemboss** server will recognise any new applications when a new client connection is made.

3.4   Installing as a non-authenticating client-server

A non-authenticating server will allow anyone to connect to it and to use it. Results are stored on the server under sub-directories corresponding to usernames on the client machines. Exactly the same installation procedure is used as for that of the authenticating server, i.e. you use the `install-jemboss-server.sh` script and answer the questions in the same way. However, when the following question appears:

```
Do you want Jemboss to use unix authorisation (y/n) [y]?
```

you should answer n. After doing so you will not be asked for any UID for a **Tomcat** user nor be asked for any authentication method such as PAM. The remaining prompts are the same as for the authenticating server. At the end of the installation script there is no requirement to change the ownership and privileges of the **jembossctl** application.

If you wish remote clients to be launched from a web page then follow the instructions for running `makeJNLP.sh`.

3.5   Technical details of authentication

You only need to read this section if you think you are having authentication difficulties. It explains how authentication is done depending on the method you chose. Authentication is concerned with verifying that a particular username and password is valid for a given user on your platform. The installation script gives various authentication choices and a default value. The default value is usually the one that you'll need as it is based on your previous selection of the operating system being used. A good rule of thumb, though, is if your operating system has working PAM authentication then use it.

The various other authentication choices are either platform specific (e.g. for AIX and HP-UX) or more general. Most of the time you'll be able to use to following options but see the notes below.

- Linux: `pam`
- Irix: `shadow`
- Mac OS X: `pam` (on 10.3 or above and previously noshadow)
- AIX: `aixshadow`
- HP-UX: `hpuxshadow`
- Solaris: `rshadow`

For AIX and HP-UX the above are the only possible choices.

If the above options do not work or you have a different operating system then here is a description of the possible choices which will hopefully guide you in selecting the correct one. Many will work with NIS (e.g. shadow, rshadow).

noshadow	In general this is used if your system has a /etc/passwd file which contains encrypted passwords and only a getpwnam() system call (**man getpwnam**). It is also used by early versions of Mac OS X which had an unusual authentication system.
rnoshadow	This can be used if your system has a /etc/passwd file which contains encrypted passwords and your system has the re-entrant **getpwnam_r()** call (**man getpwnam_r**).
shadow	This can be used if your system has an /etc/shadow file (or equivalent) and your /etc/passwd file therefore does not contain encrypted passwords. It is used when you have the **getspnam()** and **getpwnam()** system calls (**man getspnam, man getpwnam**).
rshadow	This is the re-entrant version of shadow password access. It can be used if your system has a /etc/shadow file (or equivalent) and your /etc/passwd file therefore does not contain encrypted passwords. If your platform has both the **getspnam_r()** and the **getpwnam_r()** re-entrant system calls then use this option (**man getspnam_r, man getpwnam_r**).
PAM	This is the preferred choice if your operating system has working PAM authentication set up. One of the above options may also work.
aixshadow	This is solely for AIX systems. It uses the **getuserpw()** and **getpwnam()** system calls.
hpuxshadow	This is solely for HP-UX systems. It uses the **getspnam()** and **getpwnam_r()** system calls. It also checks for 'trusted' setups with the **iscomsec()** system call.

If your system fits into none of the above categories then the authentication will not work. You should then contact the EMBOSS developers (emboss-bug@emboss.open-bio.org), preferably with details of which system calls your platform uses for authentication. Hopefully, it will be possible to incorporate your system's authentication scheme into the EMBOSS distribution.

3.6   Starting and stopping the Jemboss server

After installing **Jemboss** the server will be running. Two scripts are created by the installation process; one to stop the server (tomstop) and another to start the server (tomstart). These scripts are created in the EMBOSS-x.y.z/jemboss/utils/ directory and may be copied to a more convenient place.

The tomstart script must be invoked by the unprivileged username that was used for installation of the server.

The server can, of course, be started at boot time either by creating appropriate operating system start/stop scripts or by invoking a line similar to the following at the end of the boot process:

```
/bin/su - jemboss - /path/to/tomstartscript/tomstart > /dev/null 2>&1
```

where **jemboss** is the unprivileged username used for the installation.

3.6.1 Creating a web launch page for the Jemboss authenticating server

Users of the server will be invoking the client software by clicking on a web page link at your site. You therefore need to create the web page using the instructions given below. Clicking on the link will invoke **Java Web Start** on the client. This will check that the client machine is using the latest version of the **Jemboss** software and download the client software if necessary; it will certainly download the client software on its first invocation.

For your convenience, the distribution provides a script to bundle all the Java files, any SSL keystore files, the index.html file and the Jemboss.jnlp (Java Network Launching Protocol) file into one directory. It is recommended that you use the script.

The script is called makeJNLP.sh and must be run from the location where the server was installed and not from the source code directory. For example, if you installed **Jemboss** under /usr/local/emboss, as in the example above, then the script can be found in the directory /usr/local/emboss/share/EMBOSS/jemboss/utils/:

```
cd /usr/local/emboss/share/EMBOSS/jemboss/utils/
./makeJNLP.sh
```

Here is an example session using makeJNLP.sh. The answers to the prompts reflect the **Jemboss** installation example used above.

```
*** Run this script from the installed jemboss utils directory.
*** If you are using SSL the script will use the client.keystore
*** in the $JEMBOSS/resources directory to create client.jar
*** which is wrapped with the Jemboss client in Jemboss.jar.
*** Press any key to continue.
```

After pressing [Return] the client.jar file will be created and you will get the following text and prompt.

```
Create client.jar to contain client.keystore.
    The following information is used by keytool to
    create a key store....

    What is your first and last name [Unknown]?
    Alan Bleasby
```

Your response to the above should be quite straightforward; however, as it may appear in certificate security messages (see later), you may wish to use a generic name such as 'EBI Jemboss'.

```
What is the name of your organisational unit [Unknown]?
EMBOSS
```

Your organisational unit may be 'Systems' or 'Bioinformatics' or some such.

```
What is the name of your organisation [Unknown]?
EBI
```

Change as appropriate.

```
What is the name of your City or Locality [Unknown]?
Hinxton
```

Change as appropriate.

```
What is the name of your State or Province [Unknown]?
Cambridgeshire
```

Change as appropriate.

```
What is the two-letter country code for this unit [Unknown]?
UK
```

Change as appropriate.

```
Give a key password (at least 6 characters):
helloworld
```

This is a password to protect the individual private keys of the public/private key pairs generated during the signing of the Java files required by the client. A user of **Jemboss** doesn't need to know this and, frankly, neither do you really. Usual password rules should apply though, i.e. the example password isn't a good one.

```
Give a store password (at least 6 characters):
helloworld
```

This is a password to ensure the integrity of the keystore which holds the private keys. Again, a user doesn't need to know this, and usual password rules should apply.

```
Provide the validity period for the signed jars, i.e. the
number of days before they expire and new ones need to be made [90]:
90
```

The signed jar files only have a lifetime of a fixed number of days. This is a security measure. If you believe that 90 days is too short then you may increase this number. For example, as there's usually at least an annual release of EMBOSS you may wish to set the above value to 365 days.

The script has now gathered all the information it requires and the following messages will be displayed.

```
Each of the jar files will now be signed....
Signing axis.jar
```

```
Warning:
The signer certificate will expire within six months.
Signing commons-discovery.jar

[output truncated for clarity]

Signing Jemboss.jar

Warning:
The signer certificate will expire within six months.

*** The signed jar files, index.html and Jemboss.jnlp have been
*** created in the directory /usr/local/emboss/share/EMBOSS/jemboss/jnlp.
***
*** Please edit the 'codebase' line in Jemboss.jnlp.
*** Also, edit the 'Click here' line in index.html to point
*** href at Jemboss.jnlp.
*** The 'jnlp' directory will then need to be added to your HTTP
*** server configuration file or moved into the www data
*** directories.
***
*** For your http server to recognise the jnlp application, the
*** following line needs to be added to the mime.types file:
*** application/x-java-jnlp-file jnlp
```

The above text gives a short description of what steps you now need to take: here is a more full description. The makeJNLP.sh script will have created a new directory namely, using the above example, /usr/local/emboss/share/EMBOSS/jemboss/jnlp. This directory contains all the **Jemboss** files required by your http server.

```
% ls
axis.jar                    Jemboss_logo_large.gif      sjaxrpc.jar
commons-discovery.jar       JembossPrintAlignment.jar   sJemboss.jar
commons-logging.jar         jembossstore                sJembossPrintAlignment.jar
grout.jar                   saaj.jar                    ssaaj.jar
index.html                  saxis.jar                   swsdl4j.jar
jakarta-regexp-1.2.jar      scommons-discovery.jar      wsdl4j.jar
jalview.jar                 scommons-logging.jar
jaxrpc.jar                  sgrout.jar
Jemboss.jar                 sjakarta-regexp-1.2.jar
Jemboss.jnlp                sjalview.jar
```

First, edit the index.html file in the above directory. A little way down the file you'll find the following lines.

```
<b><font size=+1 color="#FF0000">Click here to <a
href="http://localhost/Jemboss/Jemboss.jnlp">
LAUNCH JEMBOSS</a></font></b></li>
```

You should replace 'localhost' with the IP address of your server. In our example installation this becomes:

```
<b><font size=+1 color="#FF0000">Click here to <a
href="http://192.168.8.11/Jemboss/Jemboss.jnlp">
LAUNCH JEMBOSS</a></font></b></li>
```

The location Jemboss will be created as an alias. This is explained shortly.

Second, you need to edit the Jemboss.jnlp file. Near the top of this file you will see the following lines.

```
spec="1.0+"
codebase="http://EDIT"                    .
href="Jemboss.jnlp">
```

You need to edit the codebase line to point to the http directory. For our example installation these become:

```
spec="1.0+"
codebase="http://192.168.8.11/Jemboss/"
href="Jemboss.jnlp" >
```

Third, you need to add an alias called 'Jemboss' to your httpd server configuration file. For apache servers this file is usually ... /conf/httpd.conf where the three dots refer to the path to the httpd configuration files – this differs from system to system. For example, under Fedora Linux, the file is located at /etc/httpd/conf/httpd.conf. Using the given installation example, you need to add the following line at the bottom of the file:

```
Alias Jemboss /usr/local/emboss/share/EMBOSS/jemboss/jnlp
```

This ensures that, when a user points his browser to http://yourserver/Jemboss, then he will be accessing the files in the above directory.

Finally, your httpd server may not know how to handle jnlp files correctly. Find the file mime.types associated with your httpd server. The location varies from system to system. For example, under Fedora Linux, the file is /etc/mime.types. Make sure that this file contains the following declaration:

```
application/x-java-jnlp-file jnlp
```

If there is a jnlp declaration in the mime.types file that looks different to the above then it is usually advisable to comment out that line and add the one given above.

> **Note**
>
> After editing the file **/etc/mime.types** you should restart your httpd server for the changes to take effect.

Your **Jemboss** server is now ready to use.

For those interested in technical details, the following describes what the **makeJNLP.sh** command really does. The content of the jnlp is:

```
<?xml version="1.0" encoding="UTF-8"?>
<jnlp
    spec="1.0+"
    codebase="jnlp_axis"
    href="Jemboss.jnlp">
     <information>
        <offline-allowed />
     </information>
     <information>
        <title>Jemboss</title>
        <vendor>HGMP-RC</vendor>
        <homepage href="/Jemboss/"/>
        <description>Jemboss</description>
        <description kind="short">User interface to EMBOSS.
        </description>
        <icon href="../Jemboss_logo_large.gif"/>
     </information>
     <security>
        <all-permissions/>
     </security>
     <resources>
        <j2se version="1.3+"/>
         <jar href="saxis.jar"/>
         <jar href="scommons-logging.jar"/>
         <jar href="scommons-discovery.jar"/>
         <jar href="sJemboss.jar" main="true"/>
         <jar href="sjakarta-regexp-1.2.jar"/>
         <jar href="sjalview.jar" download="lazy"/>
         <jar href="sJembossPrintAlignment.jar"/>
         <jar href="sjcert.jar"/>
         <jar href="sjnet.jar"/>
         <jar href="sjsse.jar"/>
         <jar href="ssaaj.jar"/>
     </resources>
     <application-desc main-class="org.emboss.jemboss.Jemboss"/>
</jnlp>
```

As you can see, it mentions several jar files. Most of them are signed versions of jar files from other projects e.g. saxis.jar is the signed version of axis.jar and is used by the (s) Jemboss.jar file for SOAP procedures.

The script first makes the Jemboss.jar file, using the makeJar.csh script in the same directory. The Jemboss.jar file contains the **Jemboss** Java class files, images and sundry files including the jemboss.properties file described later.

It then creates a keystore using a standard public/private encryption key methodology. The Java **keytool** command is used for this, i.e.:

```
keytool -genkey -alias signFiles -keypass yourKeyPass -keystore jemboss-
store -storepass yourStorePass
```

Finally, **jarsigner** is used to sign all the jar files. The 's' prefix is used when the signed files are created, e.g.:

```
jarsigner -keystore jembossstore -signedjar sJemboss.jar Jemboss.jar
signFiles
```

```
Enter Passphrase for keystore: <yourKeyPass>
Enter key password for signFiles: <yourStorePass>
```

One reason to bear these technical details in mind is that if, for whatever reason, you decide to edit the jemboss.properties file then you need to rerun the makeJNLP.sh procedure.

3.7   What to do if your certificates expire

There are two kinds of certificate that can run out:

- The SSL certificate: this will eventually happen if you selected SSL encryption when running the install-jemboss-server.sh script.
- The client jar file certificates: this will happen after the time limit you specified when running the makeJNLP.sh script.

If your SSL certificate runs out then you will need to create a new one using the keys.sh script provided in the utils directory. This will be /usr/local/emboss/share/ EMBOSS/Jemboss/utils in the example above. If you are launching **Jemboss** using a web page then you must run makeJNLP.sh again as well. This is because the SSL certificate is bundled with the signed jar files.

If, however, only the jar file certificates have expired then you only need to run makeJNLP.sh again.

Whenever any of these certificates are replaced you should stop and restart the server using the tomstop and tomstart scripts.

Should you need to use keys.sh, here is a sample session:

```
%cd /usr/local/emboss/share/EMBOSS/Jemboss/utils
   %mkdir tmp
   %cd tmp
   %../keys.sh

Enter where to store the keys and certificates:
.
```

The '.' specifies the current directory.

```
Enter your surname:
Bleasby

Enter a password to use to create the keys with
(at least 6 characters):
helloworld
```

This should be the same password you used when running the install-jemboss-server.sh script. If you have forgotten what you used you can see it in the **Tomcat** conf/ server.xml file. Alternatively, use a new password and replace the one in the server.xml file.

```
Provide the validity period for these certificates, i.e. the
number of days before they expire and new ones need to be made [90]:
90
```

The meaning of the above is obvious. The script now has all it needs and prints the following text to the screen.

```
Certificate stored in file <./server.cer>
Certificate stored in file <./client.cer>
*********IMPORTING
Certificate was added to keystore
[Storing ./client.keystore]
*********IMPORTING
Certificate was added to keystore
[Storing ./server.keystore]
```

The following files will have been created.

```
% ls
client.cer client.keystore server.cer server.keystore
```

These files should be copied to the **Jemboss** resources directory. For example given this would be done by typing:

```
% cp *.cer *.keystore /usr/local/emboss/share/EMBOSS/jemboss/resources
```

3.8   The jemboss.properties file

The **Jemboss** client and server use the file jemboss.properties as their primary source of setup information. The client uses this file to determine the name and location of the server; the server uses this file to define where EMBOSS is installed. If you used the install-jemboss-server.sh script to install **Jemboss** in client-server mode or if you use the standalone **Jemboss** produced by an EMBOSS **make install** then this will have been automatically generated (.../share/EMBOSS/jemboss/resource/jemboss. properties) so you can skip this section. The information here is provided for completeness and for anyone wishing to customise the file.

Here is an example jemboss.properties file:

```
jemboss.server=true
user.auth=true
server.public=https://localhost:8443/axis/services
server.private=https://localhost:8443/axis/services
service.public=JembossAuthServer
service.private=JembossAuthServer
plplot=/usr/local/emboss/share/EMBOSS/
embossData=/usr/local/emboss/share/EMBOSS/data/
embossBin=/usr/local/emboss/bin/
embossPath=/usr/bin/:/bin:/packages/clustal/:/packages/primer3/bin:
acdDirToParse=/usr/local/emboss/share/EMBOSS/acd/
embossURL=http://emboss.open-bio.org/rel/dev/apps/
```

The first line (`user.auth=true`) indicates that the server is expecting the user to login with a username and password.

The second line (`jemboss.server=true`) indicates you are using the Java JembossServer methods.

If you are using this file as a template then change 'localhost' in the above server names to be your IP address. The next lines (`service.public` and `service.private`) should be set to either JembossServer or JembossAuthServer.

This file also specifies the paths to EMBOSS-related directories. You should ensure that you end directory names with a forward slash (/).

Auto-generated `jemboss.properties` files can be found in the `.../share/EMBOSS/jemboss/resource` directory. **Jemboss** clients will read the properties file via the Java classpath. However, you may create a customised version of `jemboss.properties` in your home directory. If you do so then that file will be used by default.

jemboss.properties used by the client
An example `jemboss.properties` file that can be used by a **Jemboss** client to connect to a server that doesn't require user authorisation:

```
user.auth=false
jemboss.server=true
server.public=http://localhost:8080/axis/services
server.private=http://localhost:8080/axis/services
service.public=JembossServer
service.private=JembossServer
embossURL=http://emboss.open-bio.org/rel/dev/apps/
```

jemboss.properties used by the server
An example `jemboss.properties` file that can be used by a **Jemboss** server that doesn't require user authorisation:

```
user.auth=false
plplot=/usr/local/emboss/share/EMBOSS/
embossData=/usr/local/emboss/share/EMBOSS/data/
embossBin=/usr/local/emboss/bin/
embossPath=/usr/bin/:/bin:/packages/clustal/:/packages/primer3/bin:
acdDirToParse=/usr/local/emboss/share/EMBOSS/acd/
```

jemboss.properties definitions
To indicate that the server is not (if false) expecting any user authentication:

```
user.auth=false
```

To indicate whether **Jemboss** should run in client mode, connecting to a **Jemboss** server (as opposed to standalone mode):

```
jemboss.server=true
```

The name of the public server; this will look like a URL:

```
server.public
```

The name of the private server; this will look like a URL:

```
server.private
```

The name of the service to connect to on the public server:

```
service.public
```

The name of the service to connect to on the private server:

```
service.private
```

Plplot graphics library directory (where the `*.fnt` files are):

```
plplot
```

Location of EMBOSS data directory:

```
embossData
```

EMBOSS binary directory:

```
embossBin
```

Path used in the environment to run the EMBOSS applications
There are applications that may need to be added to the path such as **clustalw** and
primer3_core. Also, if emboss.default is set up to use SRS to retrieve sequences from
databases, the path to **getz** should be added:

```
embossPath
```

Location of the acd directory in the EMBOSS installation:

```
acdDirToParse
```

URL for emboss application documentation (e.g. `http://emboss.open-bio.org/
rel/dev/apps/`):

```
embossURL
```

It is possible for **Jemboss** to get the application help from the **Jemboss** server. This can be done using a blank entry for this property in `jemboss.properties`, i.e.:

```
embossURL=
```

It will then not use a URL but get the **Jemboss** server to run the EMBOSS **tfm** application to retrieve the help.

Other environment variables for running emboss:

```
embossEnvironment
```

For example:

```
embossEnvironment=OMP_NUM_THREADS=4 LD_LIBRARY_PATH=/usr/local/lib
```

> **Note**
>
> If you change any parameters relating to the client and you are using a web page launching method then you will need to rerun `makeJNLP.sh`.

3.9 Setting up Jemboss to use batch queuing software

It is possible to set up the **Jemboss** server to submit jobs to a network queuing system, but this requires editing the **Jemboss** source code before installation. Batch jobs submitted through **Jemboss** will be sent to a queue rather than just run in the background. The queuing software just needs to be able to take and run script files. Here are details of the server code needed, although further changes may be required, depending on the batch queue system used:

```
Edit EMBOSS-x.y.z/jemboss/org/emboss/jemboss/server/JembossAuthServer.java
```

First comment out this line (which is used to submit batch jobs as background processes):

```
aj.forkBatch(userName,passwd,environ,embossCommand,project);
```

Then uncomment this line to call (`runAsBatch()`):

```
runAsBatch(aj,userName,passwd,project,embossCommand)
```

In the method `runAsBatch()` the following line will need to be changed. The `batchQueue.sh` is a script you will need to create to submit the emboss script (i.e. `project/.scriptfile`) to the batch queue system:

```
//EDIT batchCommand
String batchCommand = "/bin/batchQueue.sh " + project +
                                 "/.scriptfile ";
```

The `.scriptfile` contains the script (environment variables and emboss command) for the queuing system to run. This file is created in the users area under a new project directory for that EMBOSS run. The next line:

```
lfork = aj.forkEmboss(userName,passwd,environ,
                       batchCommand,project);
```

should not need changing. This line calls the emboss AJAX library to submit the batch command (as that user) to the queuing software.

3.10 Setting up the clients

The clients can use any hardware and operating system that can support SUN or OpenJDK Java, and possesses moderate graphics capabilities. Once Java has been loaded onto the machine, the user just needs to point his browser at the web page set up above and click on the launch link. This should fire up Java Web Start. If not, then a box will appear asking which application should be used for the `Jemboss.jnlp` file. On UNIX systems this will be `[path-to-java]/bin/javaws`.

Note

64-bit SUN Java has only recently started supporting Java Web Start. Before then, clients using such hardware would be configured to use the 32-bit standard Java which works quite happily on 64-bit architectures.

3.11 Troubleshooting Jemboss installation

If a **Jemboss** server was working but has stopped doing so then stop the server using the `tomstop` script, check that the server process has stopped and kill it if not, and then restart the server using the `tomstart` script. Allow 30 seconds for the server to stabilise.

If the client-server installation fails, and if the cause is not immediately obvious, then two log files may help diagnose the problem. Using the installation example above these are:

```
/m1/SOAPRESULTS/Jemboss_error.log
/usr/local/tomcat/logs/catalina.out
```

The most common problems are:

- Failure to set the permissions of the `jembossctl` program correctly (client-server). This should be owned by root with the `setuid` bit set.
- Failure to set the permissions of the results directory correctly. This should be world readable and writeable with the sticky bit set.
- Using a very new version of **Tomcat** or **Axis** that **Jemboss** has not yet been modified to work with.
- EMBOSS binaries not working owing to missing library dependencies. You should check that the EMBOSS binaries work by invoking one of them directly. In the example installation you should try typing:

```
/usr/local/emboss/bin/embossversion
```

Jemboss will not work if the EMBOSS binaries don't.

- Ports blocked by firewalls or already in use by other applications.

Important

Note that if you have to try reinstalling **Jemboss** then you *must* start from scratch. We'll repeat that; you must start from scratch! Installation of **Jemboss** modifies things in most directories such that they cannot easily be set back to their default contents by any other means than deletion and re-extraction. So, delete:

- The EMBOSS source code directory
- The installation directory
- The tomcat directory
- The axis directory

then re-extract them from their tarballs.

3.12 Jemboss installation: platform-specific concerns

3.12.1 Mac OS X

These instructions refer to the 10.6 and 10.5 versions of Mac OS X although similar principles apply to 10.3 and 10.4.

Note

Also note that the EMBOSS developers can only realistically support the package on a virgin Mac OS X installation using the installation methods described above. If you are having installation problems with any form of **Jemboss** provided by another project (e.g. MacPorts or FINK) then you should contact them for support.

3.12.1.1   General prerequisites

First, make sure you have acted on the prerequisites section for this operating system in the EMBOSS chapter. You will need Xcode, **X11**, **X11SDK** and PNG graphics. On top of that you will also need to have installed Java and the Java SDK. Depending on your version of Mac OS X, not all of these may be installed by default so insert your Mac OS X DVD (or look at the Xcode installation options) if necessary and do a *custom* installation and ensure that relevant check boxes are ticked.

Note

With later versions of Mac OS X, the Apple Xcode software should detect whether you have **X11** installed and, if so, will install the X11SDK files. To be really sure of getting the latest versions of these packages then you will have to register as a developer at http://connect. apple.com. The process is simple, painless and free.

3.12.1.2   When running the script

The Java location prompt should be answered with:

```
/System/Library/Frameworks/JavaVM.framework/Home
```

You should select PAM authentication for the latest versions of Mac OS X.

When asked whether you wish to add any more configuration options, be sure to specify the location of your PNG installation e.g.

```
--with-pngdriver=/usr/local
```

Prior to asking you for a password, the script may produce a warning message saying that it doesn't know what to set LD_PRELOAD to in the tomstart script. This warning can be ignored. For completeness you may wish to edit the tomstart script and change /usr/lib/libpam.so to /usr/lib/libpam.dylib.

Prior to automatic deployment you may be requested to edit the java.security file. This is located at:

```
/System/Library/Frameworks/JavaVM.framework/Versions/version/Home/lib/
security/java.security
```

Where *version* is your current version of Java e.g. 1.6.0. Typing:

```
java -version
```

in a terminal window will allow you to correctly choose which version to enter in the above file specification.

You may, however, notice that the java.security file already contains the required SSL provider line in which case no action needs to be taken.

3.12.2 IRIX

3.12.2.1 General prerequisites

Follow the prerequisites from the EMBOSS installation chapter. In addition you will need to download and install Java for IRIX from the SGI website. The current version is 1.4.1. It installs into /usr/java2. As this is an older version of Java then you will need to download and install the **Tomcat** compatibility binaries on top of the **Tomcat** base binaries before running the **Jemboss** installation script.

3.12.2.2 When running the script

You should set the CC environment variable before running the script, e.g.:

```
setenv CC cc
```

or

```
export CC=cc
```

depending on your shell.

When prompted for any extra configuration options, enter both the PNG and SGI compiler specification here (see Chapter 1, *Building EMBOSS*). For example:

```
--with-pngdriver=/usr/local -with-sgiabi=n32m4
```

Databases

4

4.1 General database configuration

4.1.1 Sequence database support

EMBOSS provides excellent database support. All the common sequence formats you are likely to come across are supported. See the *EMBOSS User's Guide*.

A variety of indexing and access methods are supported. For example, *EMBL* entries can be read from :

- A non-indexed EMBL-format flatfile held locally
- Original *EMBL* flatfiles using the CD-ROM, Staden or EMBOSS indexes
- Original *EMBL* flatfiles using local SRS indexes
- A file indexed for use with **BLAST** version 2 indexes
- GCG database format
- A query to the EMBL-EBI DBFETCH service
- A query to the EMBL-EBI web server
- A query to the Entrez web server
- A query to any MRS web server
- A query to any SRS web server (local or remote)
- A relational database such as Sybase or Oracle by calling a local application.

Databases can be held locally and both indexed and non-indexed local files are supported. Tools for database indexing (Section 4.5, 'Database indexing') are provided. One is a variation on the emblcd system, the other uses an updatable tree. They provide rapid access to single sequences and rapid queries of flatfile databases. The **dbi*** indexing applications assume that you have one or both of ID and accession number in each record and that they are unique for the whole database index, whereas the **dbx*** applications can handle non-unique (duplicate) IDs and source files >2 Gb in size. Use of the **dbx*** indexing applications is preferred.

EMBOSS also provides methods for retrieving sequences via the WWW. If sequences on a server are in a format unknown to EMBOSS, it might be possible to specify they are converted to FASTA format before they are served. There are three methods for interaction with a local SRS installation or SRS on a remote public server. SRS queries can be made not only by ID and accession number, but also (depending on the way a database has been indexed) on words in the description line, sequence version (or GI numbers), keywords or organism names.

Specialised access methods are provided for databases served by MRS, NCBI's Entrez and EMBL-EBI's dbfetch servers

For more general access through web servers, the url access method allows a database to be defined as a URL into which a user-specified ID is inserted.

For other non-flatfile databases or flatfile databases in formats not currently supported by EMBOSS, it is possible to configure an external application to retrieve sequences.

4.1.1.1 Query levels, access methods and attributes

There are three basic levels of query:

entry   A single entry specified by database ID or accession number is retrieved

query   One or more entries matching a wildcard string in the Uniform Sequence Address (USA; see the *EMBOSS User's Guide*) are retrieved (this can be slow for some methods)

all   All entries are read sequentially from a database

One or more query levels may be specified for each database configuration.

There are many methods (Section 4.3, 'Database access methods') for accessing databases. The available methods depend on the query level, i.e. whether a single entry, a wildcard-specified set of entries or all of the database entries are to be retrieved. For example, a web server might be suitable for retrieving a single or few entries but probably, quite sensibly, will not allow an entire database to be retrieved over the internet. In contrast, a flatfile database with no index is often (depending upon its size) only useful for reading all the entries sequentially ('all' retrieval level).

A database can be defined with a single retrieval method using the method attribute. Alternatively, multiple methods may be defined, depending on which type (entry, query, all) of access is required. The attributes methodentry, methodquery and methodall are used for this. This would be essential in the cases described above, to access the database in the different locations.

In addition, each access method needs to know something about the database. What is needed will be different for each method, although there is, of course, much overlap between them. This information is specified by using the 'key: value' attributes. The required attributes depend on the access method and the query level.

Database key: value attributes and access methods (Section 4.3, 'Database access methods') are described below.

4.1.2 Configuring EMBOSS to work with databases

Every database you intend to use must be defined in one of the EMBOSS configuration files:

```
emboss.default
.embossrc
```

93

emboss.default is kept in the top-level EMBOSS directory (e.g. /usr/local/ emboss/share/EMBOSS/emboss.default) and is used for defining site-wide databases. In contrast, .embossrc lives in your home directory and is used for defining your own databases or, for example, testing database definitions before adding them to the site-wide emboss.default file.

Each database is configured using a database definition. The generalised form is:

```
DBNAME DatabaseName
[
     key: value
     key: value
     key: value
     key: value
]
```

DBNAME, which is usually shortened to DB, is followed by the database name (*DatabaseName*) then a set of *key: value* attributes that specify that database. The *key: value* attributes are all enclosed by a pair of square brackets.

The *key: value* pairs are the configuration options and must contain:

- A description of the access method (using method:) or one or more of:

```
methodsingle:
methodquery:
methodall:
```

- A description of the original format of the sequences (using format:).

Additional *key: value* pairs might be required depending on the access methods. Others are optional.

As an illustration, to set up direct access to the *EMBL* and *SwissProt* test databases distributed with EMBOSS, your emboss.default or .embossrc file should look something like this:

```
DB embl
[
type:     "N"
method:   "direct"
format:   "embl"
dir:      "/home/auser/EMBOSS-6.2.0/test/embl/"
file:     "*.dat"
comment: "Test EMBL in EMBOSS distribution"
]

DB swissprot
[
type:     "P"
method:   "direct"
format:   "swiss"
```

```
dir:      "/home/auser/EMBOSS-6.2.0/test/swiss/"
file:     "seq.dat"
comment: "Test Swissprot in EMBOSS distribution"
]
```

Or to set up access to the *EMBL* and *SwissProt* databases via SRS at the EMBL-EBI, your emboss.default or .embossrc file should look like this:

```
DB swissprot
[
type:     "P"
method:   "srswww"
format:   "swiss"
url:      "http://srs.ebi.ac.uk/srsbin/cgi-bin/wgetz"
comment: "Swissprot via EBI SRS"
]

DB embl
[
type:     "N"
method:   "srswww"
format:   "embl"
url:      "http://srs.ebi.ac.uk/srsbin/cgi-bin/wgetz"
comment: "EMBL via EBI SRS"
]
```

4.1.3   Example database definition file (emboss.default.template)

An emboss.default.template file is provided in the EMBOSS distribution. As its name suggests, it gives examples of some of the possible database definitions supported by EMBOSS (see the next section). An excerpt of the emboss.default.template file is show below:

```
#SET emboss_tempdata path_to_directory_$EMBOSS/test

# Logfile – set this to a file that any user can append to
# and EMBOSS applications will automatically write log information

#SET emboss_logfile /packages/emboss/emboss/log

# pir (cytochrome C plus first entries in other divisions)
# ===

DB tpir [
    type: P
    dir: $emboss_tempdata/pir
    method: gcg
    file: pir*.seq
    format: nbrf
    fields: "des org key"
    comment: "PIR in 4 files in GCG format indexed by dbigcg"
]
```

```
# Genbank (Remote access to an MRS server)
# =======

DB genbank [
    type: N
    methodentry: mrs3
    format: genbank
    dbalias: "genbank_release"
    url: "http://mrs.cmbi.ru.nl/mrs-3/plain.do"
    comment: "GenBank IDs via MRS"
]

# genbank (the first few entries from several sub-section files)
# =======

DB tgenbank [
    type: N
    dir: $emboss_tempdata/genbank
    method: emblcd
    format: genbank
    release: 01
    fields: "sv des org key"
    comment: "GenBank native format indexed by dbiflat"
]
```

4.1.4   Test databases

To see how databases are set up under EMBOSS, you should look at the configurations for the test databases included in the EMBOSS distribution. The EMBOSS developers use these databases to test database indexing and sequence reading. They also contain the sequences that are used in the usage examples for the applications (see the application documentation online or by running **tfm**). They include:

- test/data (emrod (DNA) and swnew (protein) are in **BLAST** format)
- test/embl (*.dat for *EMBL* format, .ref and .seq for gcg format)
- test/pir (.ref and .seq for nbrf format)
- test/swiss (.dat for swissprot format, 1 file)
- test/swnew (.dat for swissprot format, 3 files)
- test/wormpep (wormpep is in FASTA and **BLAST** format)

The template file (emboss.default.template) in the EMBOSS distribution (e.g. /usr/local/emboss/share/EMBOSS/emboss.default.template) contains configurations for all the test databases. You can use emboss.default.template as a template for entries in your own emboss.default file. For any database definitions you use, change the definition of emboss_tempdata to point to your test directory and uncomment the line. You'll then be able to use the test databases as "tembl", "tsw" and so on.

One of the first things an EMBOSS application does when it runs is to read in the installed emboss.default (and then the ~/.embossrc file, if it exists). This means that any changes to these definition files take effect as soon as they are made.

For example, change:

```
# swissprot (Puffer fish entries)
# =========

DB tsw [ type: P dir: $emboss_tempdata/swiss
    method: emblcd format: swiss release: 36
    fields: "sv des org key"
    comment: "Swissprot native format with EMBL CD-ROM index" ]
```

to

```
# swissprot (Puffer fish entries)
# =========

DB tsw [ type: P dir: /home/auser/EMBOSS-6.2.0/test/swiss
    method: emblcd format: swiss release: 36
    fields: "sv des org key"
    comment: "Swissprot native format with EMBL CD-ROM index" ]
```

Alternatively, to get all the test databases supported, rename or copy emboss.default. template to emboss.default and edit the file as follows. This line:

```
# SET emboss_tempdata path_to_directory_$EMBOSS/test
```

must be uncommented and the definition changed to the directory where the databases are installed. In the following example this is /usr/local/share/EMBOSS/test. For example:

```
SET emboss_tempdata /usr/local/share/EMBOSS/test
# or
SET emboss_tempdata /home/auser/workspace/emboss/emboss/test/
# or something else
```

> **Note**
>
> The directory where the test databases are installed can be changed with --prefix when you configure EMBOSS.

4.1.5   Testing your database definitions

Having defined your databases (see Section 4.1, 'General database configuration'), you can run **showdb -full** and you should see them all appear in the list of databases. If the message Warning: Bad database definition is generated or if a database doesn't appear then something is seriously wrong with your definition. Go back to it and check things. Common mistakes include:

1. Have you left off the terminal square bracket] ?
2. Did you leave out a colon character : in an attribute?
3. Have you forgotten to put in the closing quotes around some text?
4. Is the emboss.default file world-readable?

If **showdb** displays your database, check that all of your required access methods are listed as OK. If something is not OK then another access method might be required.

Just because **showdb** finds a database definition does *not* mean the database is working correctly: **showdb** does not attempt to extract any entries from your database. Therefore you should try extracting one or more known entries from the database using **seqret**. If you get errors, you should check that the database is set up correctly and defined correctly. Things to check include:

- Are the data files and indexes world-readable?
- If using method: emblcd, gcg, blast or emboss did you index the data files?
- If using app: is the application in your PATH?
- If using app: is the PATH specified correctly?
- If using app: is the application world-executable?
- If using url: or srswww is the server up?
- If using url: or srswww is the server URL correct?
- Are file: wildcards specified correctly?
- Are directory: paths specified correctly?
- Have you put the files there yet?
- If using any SRS method, did you use dbalias:?
- If using any SRS method, check the dbalias: name in the SRS server.
- If accessing by SV (GI), DES, KEY or ORG, did you remember to specify these when you indexed the database?
- If accessing by SV (GI), DES, KEY or ORG, did you specify fields:?
- Take another look at the format. Is that really fasta, or is it ncbi?
- Do you have duplicate entries? The **dbi*** program indices must have unique entry names.

4.2 Database attributes

4.2.1 Introduction

There are a few things to consider when specifying attributes for a database:

- Each database must have attributes that specify what it is and how to access it. This information is given as a set of pairs of key: value attributes. These attributes are held in the DB definition structure (see above).
- The key: value pairs in a DB structure can be specified either on separate lines or separated by spaces on the same line.

- If the `value` part of the attribute contains spaces then it should be quoted to prevent it being prematurely terminated at the first space. For example, `key: "value with many words in"`.
- The minimum set of attribute keys are `method:` and `format:` – these two are mandatory. It is also typical (but not mandatory) to specify the `type:` attribute.
- Some forms of `method:` require subsidiary attributes giving further information on how to access the data.

The available attributes are described below.

Table 4.1 Attributes used to specify a database

Key	Value	Description
method methodall methodentry methodquery	srs srsfasta srswww url app external direct emblcd emboss entrez gcg embossgcg blast dbfetch mrs direct	Specifies the method used to access the database.
format formatentry formatquery formatall	A valid sequence format name (see the *EMBOSS User's Guide*)	Specifies what sequence format to expect when reading entries from the database.
type	N or P	Specifies whether the database is nucleic or protein.
fields	One or more of: sv, des, org, key	Specifies which search fields have been indexed and are available for searching with.
directory	Any valid directory path	Specifies the directory of files that have been specified with the `filename:` attribute. It also specifies the default directory of indexes and files produced by the **dbi*** and **dbx*** indexing programs (see `indexdirectory:`).
filename	A filename (may be wildcarded) or list of filenames	Specifies the sequence file(s) to read in when accessing the database.

Table 4.1 (cont.)

Key	Value	Description
exclude	A filename (may be wildcarded) or list of filenames	This is used to exclude a subset of files from consideration.
indexdirectory	Any valid directory path	Specifies the directory of index files (produced by the **dbi*** and **dbx*** programs) if this is different to the directory specified by directory:.
url	Any valid URL	Specifies the URL to use when getting sequences from remote web sites.
httpversion	1.0 or 1.1	Specifies the HTTP version to be used. Version 1.0 transmits the results in one block. Version 1.1 chunks data and is preferred for large data transfers. The default is 1.1.
proxy	*host:port*	In the access methods srswww and url, you can specify a proxy *host* and *port* to use when accessing the URL. If a proxy is globally defined, it can be bypassed for any database by specifying ":" as an empty value.
app appentry appquery appall	Any script or program name	Specifies the name or command line of an external (i.e. non-EMBOSS) program or script (application) that should be run to extract the sequence from the database.
dbalias	The true name of a database	This is used to specify the name of a database at a (e.g. SRS) site where the name differs from the name that given as the DBNAME. This allows the EMBOSS database definition to use another name (e.g. srsembl) or to specify a less obvious name when contacting the server (e.g. emblrelease)
caseidmatch	Used to flag databases that have case-sensitive identifiers	A boolean set to "Y" to define a database where identifiers can differ only in upper- or lower-case characters. An example is a sequence database derived from *PDB* entries where the chain identifiers 'a' and 'A' are not the same.
hasaccession	Used to flag databases that do not have access by accession number	A boolean set to "N" to define a database with no accession numbers (e.g. *PDB* used as a source of sequence data).
comment	Any text	A comment, usually to describe the database.
release	Any text	This is the release number or date.

4.2.2   Description of attributes

4.2.2.1   `method`, `methodall`, `methodentry`, `methodquery`

This specifies the method used to access the database.

This field is mandatory – there must be at least one form of the `method` key specified. More than one different type of method key can be specified.

If `method:` is specified, then this is the default method covering all forms of access ('query', 'entry' or 'all'). Specific methods for the 'query', 'entry' or 'all' forms of access (i.e. `methodquery:`, `methodentry:` or `methodall:`) should be specified explicitly if you wish to have several ways of accessing the data e.g.

```
method: "emblcd"
methodall: "direct"
```

4.2.2.2   `format`, `formatentry`, `formatquery`, `formatall`

The `format:` attribute specifies what sequence format to expect when reading entries from the database.

This attribute is mandatory. If you need to specify different formats for any of the different access methods (Section 4.3, 'Database access methods'), then you may use the variants of `format:` with the suffix `entry`, `query` or `all`. An example of `format` is:

```
format: ncbi
```

4.2.2.3   `type`

This specifies whether the database is nucleic or protein.

Although it is not strictly required, it is normal to specify the type of the database as this should be known. If the type is not specified it will be determined by the EMBOSS applications when they read sequences in. (You will not get error messages when you run **showdb** as this doesn't read in sequences.) The value `Nucleotide` or `N` specifies a nucleic database, `Protein` or `P` specifies a protein database, e.g.

```
type: "Nucleotide"
```

4.2.2.4   `fields`

This specifies which search fields have been indexed and are available for searching.

It is assumed that accession number and ID name are always available when a database is set up. Depending how you set up the database, access by one or more of these fields might be possible:

sv	Sequence version or GI number
des	Description line
org	Organism's taxonomic classification
key	Keywords

The access methods srs, srsfasta and srswww allow access to these search fields. The methods emboss, emblcd and gcg may or may not have some or all of these fields indexed, depending on the parameters given to the programs **dbxflat, dbxgcg, dbiflat** and **dbigcg**. The programs **dbxfasta, dbiblast** and **dbifasta** only allow you to select any of sv, des and acc (the default). An example specification is:

```
fields: "sv des org key"
```

The use of these fields in searches is described elsewhere (see the *EMBOSS User's Guide*).

FASTA format has only an ID and a parsable description line. If accession numbers are not defined then set hasaccession: "N" to turn off the default attempt to include this field in searches. A common case is the *PDB* protein structure database when used as a source of sequences, as *PDB* has no accession number system.

4.2.2.5   directory

This specifies the directory of files that have been specified with the filename: attribute. It also specifies the directory of indexes and files produced by the **dbx*** or **dbi*** programs.

It is only required with the access methods (see Section 4.3, 'Database access methods'):

```
    emboss

    direct

    gcg

    emblcd

    blast
```

It is common to use variables (see the *EMBOSS User's Guide*) to specify part or all of the path:

```
directory: $dbdir/genomes
```

4.2.2.6   filename

This specifies the sequence file(s) to read in when accessing the database.

It is only required with the access method direct (see Section 4.3, 'Database access methods'). It may also be used with the access methods:

```
    emboss

    gcg

    emblcd

    blast
```

to indicate which files should be included back in after using the exclude: attribute to specify which indexed files should be ignored (see exclude: below). The files may be wildcarded using *. The attribute key filename: is commonly abbreviated to file: e.g.:

```
file: pir*.seq
```

A list of filenames may also be given; each name must be separated with a space or comma.

4.2.2.7 `exclude`

This is used to exclude a subset of files from consideration.

To exclude certain files, specify `exclude: *file*`. This is used in conjunction with `filename:` to specify a subset of files in a directory. `Exclude:` is checked first, then the rest of the files are included with `filename:`. The files searched are therefore: – the files in the directory specified by `directory:` – but not the `exclude:` files (if any) – but include back the `filename:` files (if any) e.g.

```
exclude: mouse.*
```

If you have indexed all of the files in the *EMBL* database, then you can specify subsets using the same set of files and indexes as:

```
DB embl [
  type:      "N"
  format:    "embl"
  method:    "emblcd"
  dir:       "/data/embl"
  comment:   "All of EMBL"
]

DB emblminus [
  type:      "N"
  format:    "embl"
  method:    "emblcd"
  dir:       "/data/embl"
  exclude:   "est*.dat"
  comment:   "EMBL without the ESTs"
]

DB emblhumest [
  type:      "N"
  format:    "embl"
  method:    "emblcd"
  dir:       "/data/embl"
  exclude:   "*.dat"
  filename:  "est_hum*.dat"
  comment:   "EMBL human ESTs"
]

DB human [
  type:      "N"
  format:    "embl"
  method:    "emblcd"
  dir:       "/data/embl"
  exclude:   "*.dat"
  filename:  "hum*.dat"
  comment:   "EMBL human"
]
```

4.2.2.8   indexdirectory

This specifies the directory of index files (produced by the **dbx*** or **dbi*** programs) if this is different to the directory specified by directory:.

For the **dbi*** applications it is sensible to hold the indexes in a different directory to the one holding the sequence database files when you have many sequence databases in the same directory. This is because the indices for every database all have the same names (acnum. hit, acnum.trg, division.lkp, etc.) and these would be overwritten if you have indexed several databases in the same directory. In this case, you should create the indices in a different directory (often but not necessarily a sub-directory) for each database. That way the index files will not become confused. These index directories can be specified using the attribute indexdirectory:, while the directory containing the sequence data files can still be specified using dir:.

It is only used with the access methods (see Section 4.3, 'Database access methods'):

```
emboss
gcg
emblcd
blast
```

It is common to use variables to specify part or all of the path. The attribute key indexdirectory: is commonly abbreviated to indexdir: e.g.:

```
indexdir: $dbdir/genomes/embl
```

4.2.2.9   url

This specifies the URL to use when retrieving sequences from remote websites.

It is only required with the access methods (see Section 4.3, 'Database access methods'):

```
srswww
url
```

The database (or the name specified in a dbalias attribute) and entry accession number (or sequence version, GI number, description, organism, or keyword) can then appended to create a functional SRS query line. Often it is only necessary to specify the remote **wgetz** application alone, e.g.:

```
url: "http://srs.ebi.ac.uk/srsbin/cgi-bin/wgetz"
```

The URL can also contain one or more instances of the character pair %s – each of these pairs are replaced by the value of the ID name when this database is accessed. Any HTML formatting will be stripped from the resulting web page, e.g.:

```
url: "http://www.ebi.ac.uk/htbin/emblfetch?%s"
# or, all on one line
url: "http://www.ncbi.nlm.nih.gov/htbin-
post/Entrez/query?db=s&form=6&dopt=g&html=no&uid=%s"
```

The URL must begin with `http://` and have a lower-case host address.

4.2.2.10   proxy

In the access methods `srswww mrs entrez dbfetch` and `url`, you can specify a proxy host and port to use when accessing the URL. For example:

```
proxy: "proxy.mydomain.com:8888"
```

If the global variable EMBOSS_PROXY is defined in the `emboss.default` file (see the *EMBOSS User's Guide*) then the attribute

```
proxy: ":"
```

will turn off proxy access for this database. This is useful if the database is on an internal server.

4.2.2.11   httpversion

In the access methods `srswww mrs entrez dbfetch` and `url`, you can specify the http protocol version to use when accessing the URL. The default version 1.1 supports delivery of results in chunks. The older 1.0 protocol can only deliver all results in one block.

For example:

```
httpversion: "1.0"
```

If the global variable EMBOSS_HTTPVERSION is defined in the `emboss.default` file (see the *EMBOSS User's Guide*) then this will set a global default for all URL-based data access. The default is 1.1.

4.2.2.12   app, appentry, appquery, appall

This specifies the command line of an external (third-party) application that should be run to extract a sequence from a database.

This application can be in the user's path or have an explicit path provided. The database and entry name will be appended to the application command as *application dbname: entry*. Both ID and accession number can be used to specify the entry. Alternatively, if the app: attribute value contains the character pair %s, it is replaced by the value of the ID name or accession number when this database is accessed.

This attribute is only required with the access method app (see Section 4.3, 'Database access methods'). If you need to specify different applications for any of the different access methods, then you may use the variants of app: with the suffix entry, query or all, e.g.:

```
app: efetch
# or
app: "getz [embl:%s]"
```

4.2.2.13 dbalias

This is used to specify the name of a database at a (e.g. SRS) site where the name differs from the DBNAME.

It is only required with the access methods (see Section 4.3, 'Database access methods'):

mrs

mrs3

srswww

srsfasta

srs

For example:

```
dbalias: emblnew
```

4.2.2.14 comment

This is a comment to describe the database.

It is displayed in **showdb**, e.g.:

```
comment: "This is my subset of refseq"
```

4.2.2.15 release

This is the release number or date.

It is displayed in **showdb**.

> **Caution**
>
> Unless you are zealous in updating **release**: values, this will rapidly become out of synch with the actual data.

The **dbx*** and **dbi*** indexing programs ask for the database name, release number and index date. These are stored in the index files. This information is *not* available to EMBOSS programs and is not reported by **showdb**. They are part of the index file formats, but EMBOSS does not currently make use of them.

```
release: "21.0 (Oct 2009)"
```

4.2.2.16 `hasaccession`

This turns off attempts to read data by accession number.

Most sequence databases follow the example set by the major public protein and nucleotide resources by providing unique accession numbers. Where these are not available the accession number search can be disabled by defining

```
hasaccession: "N"
```

4.2.2.17 `caseidmatch`

This makes identifier tests case-sensitive.

Most sequence databases attach no significance to upper- or lower-case for identifiers. In a few cases, especially in site-specific local data, there may be a distinction between two otherwise identical names. An early example was a database of sequences derived from *PDB* where the chain name 'a' or 'A' in the identifier was significant.

```
caseidmatch: "Y"
```

4.3 Database access methods

4.3.1 Introduction

The available database access methods are described below.

Table 4.2 Database access methods

Method	Scope	Comments
EMBOSS	*	Uses a b+tree index from the programs **dbxflat** (used for 'flat' files, i.e. files in their native database format) or **dbxfasta** (FASTA-format files).
EMBLCD	*	Uses an EMBLCD index from the programs **dbiflat** (used for 'flat' files, i.e. files in their native database format) or **dbifasta** (FASTA-format files).
SRS	*	This calls **getz** locally, using the **-e** switch to return whole entries in original format. Query fields supported are id, acc, gi, sv, des, org and key.
SRSFASTA	*	As for SRS, but uses **getz -d -sf fasta** to read the sequence in FASTA format. Query fields supported are id, acc, gi, sv, des, org and key.
SRSWWW	single entry	Uses a defined SRS WWW server to read a single entry. Query fields supported are id, acc, gi, sv, des, org and key.
MRS	*	This uses a defined MRS server to read a single entry. Query fields supported are id and acc.

Table 4.2 (cont.)

Method	Scope	Comments
Entrez	*	This uses Entrez at NCBI to read data. Query fields supported are id, acc, gi, sv, des, org and key. Data is returned in the original format (e.g. *GenBank*).
Dbfetch	*	This uses **Dbfetch** REST access at EBI to read data. Query fields supported are id and acc.
BLAST	*	Uses an EMBLCD index from the program **dbiblast**.
EMBOSSGCG	*	Uses a b+tree index from the program **dbxgcg** to access a database reformatted for GCG 8, 9 or 10 by GCG programs such as **embltogcg**.
GCG	*	Uses an EMBLCD index from the program **dbigcg** to access a database reformatted for GCG 8, 9 or 10 by GCG programs such as **embltogcg**.
DIRECT	all	Opens the database file(s) and returns each entry sequentially. Query fields supported are id, acc, gi, sv, des, org and key.
URL	single entry	Uses any other web server (for example the EBI's **emblfetch** or **swissfetch** queries) to return an entry.
APP EXTERNAL	*	Run an external application or a simple script which returns one/more/all entries.

4.3.2   Description of database access methods

4.3.2.1   EMBOSS

The EMBOSS index method is preferred over the older EMBLCD method. It allows for non-unique index terms e.g. non-unique IDs. It can also cope with files over 2 Gb in size.

This method uses b+tree indexes from the programs **dbxflat** (flatfiles – database native format files) or **dbxfasta** (FASTA format files). This can cope with all levels of access. Queries use the index files. Reading all entries uses the list of files in the [database].ent file and opens each in turn.

Supports queries by

    id
    acc
    sv
    key
    org
    des

(*Not* by key and org if the database was indexed by **dbxfasta** as these cannot be found in the FASTA format description line.)

The directory containing the sequence files and indexes to be read must be specified using the directory: attribute. If the indexes are in a directory other than the one containing the sequence files, then the index directory can be explicitly set using the `indexdirectory:` attribute.

The available fields should be specified using the `fields:` attribute if more than just the default ID name and accession number fields have been indexed. As these indexes allow non-unique IDs then each of the fields may return a list of matches, i.e. type `query` is used throughout.

For example:

```
DB mydb [
    type: N
    method: emboss
    format: embl
    fields: "sv des org key"
    directory: /data/embl
]
```

The EMBOSS b+tree index files include the filenames indexed by **dbxflat** or **dbxfasta**. You can use the `file:` and `exclude:` attributes to create file-specific subsets from a single index.

4.3.2.2   EMBLCD

Uses an EMBLCD index from the programs **dbiflat** (flatfiles – database native format files) or **dbifasta** (FASTA format files). This can cope with all levels of access. Queries use the index files. Reading all entries uses the list of files in the `division.lkp` file and opens each in turn.

Supports queries by

    id

    acc

    sv

    key

    org

    des

(*Not* by key and org if the database was indexed by **dbifasta** because there is no way to find these in the FASTA format description line.)

The directory containing the sequence files and indices to be read must be specified using the `directory:` attribute. If the indices are in a directory other than the one containing the sequence files, then the index directory can be explicitly set using the `indexdirectory:` attribute.

The available fields should be specified using the `fields:` attribute if more than just the default ID name and accession number fields have been indexed. A wildcard search for unique fields (`id` or `sv`), or any search for `acc`, `des`, `org` or `key` is of type `query` and returns a list of entries. A search for a single `id` or `sv` is of type `entry` and will find the first match in the index and assume no other matches. The ID has to be unique in an EMBLCD database.

For example:

```
DB mydb [
   type: N
   method: emblcd
   format: embl
   fields: "sv des org key"
   directory: /data/embl
]
```

The EMBLCD index files include the filenames indexed by **dbiflat** or **dbifasta**. You can use the file: and exclude: attributes to create file-specific subsets from a single index. Use of the indexdir: attribute is common, allowing index files to be in a different directory from the source flat files.

4.3.2.3   SRS

This requires a local installation of SRS. This calls **getz** locally, using the -e switch to return whole entries in the original format. It is expected that **getz** is in the path.

Supports queries by

```
id
acc
sv
key
org
des
```

If the SRS server has a different name for this database than that specified as the DBNAME, then you must specify it using the dbalias: attribute.

EMBOSS expects the SRS local access program to be called **getz**, but you can explicitly override this using the app: attribute. This can be used to call **getz** using its explicit path, rather than relying on **getz** being in the path.

Database definitions using method: srs should also specify methodall: direct plus directory: and file: for reading all entries directly. This is much faster than using **getz** to read and format all entries (unless the database is very small).

For example:

```
DB mydb [
   type: N
   format: embl
   method: srs
   dbalias: embl
   fields: "sv des org key"

#  define 'all' access method
   methodall: direct
   directory: /data/embl
   file: *.seq
]
```

As SRS returns the results using **getz -e**; the format should match the format of the original data. For some formats this might be problematic (*PIR* for example). In that case you can consider using SRSFASTA although this will lose information that is not included in the FASTA format SRS output.

4.3.2.4 SRSFASTA

As for SRS, but uses:

```
getz -d -sf fasta
```

to read the sequence in FASTA format. It is used for databases like dbEST.reports where EMBOSS does not understand the entry format but SRS can convert it to FASTA. As the database format is not understood by EMBOSS, a search of the entire database would be forced to use **getz** to convert each entry, which would be slow.

Supports queries by

```
id
acc
sv
key
org
des
```

If the SRS server has a different name for this database from that specified as the DBNAME, then you must specify it using the dbalias: attribute.

EMBOSS expects the SRS local access program to be called **getz**, but you can explicitly override this using the app: attribute. This can be used to call **getz** using its explicit path, rather than relying on **getz** being in the path.

Database definitions need to specify methodall: direct plus directory: and file: to read all entries directly. This is much faster than using **getz** to read and format all entries.

For example:

```
DB mydb [
    type: N
    format: fasta
    method: srsfasta
    dbalias: embl
    fields: "sv des org key"

#   define 'all' access method
    methodall: direct
    directory: /data/embl
    file: *.seq
]
```

4.3.2.5   SRSWWW

Uses a defined SRS web server to read a single entry. This can be useful, for example, to get the *GenBank* version of an *EMBL* entry. Wildcard entry names are not recommended because SRS servers are not intended to return large numbers of entries.

Supports queries by

```
id
acc
sv
key
org
des
```

If the SRS server has a different name for this database from that specified as the DBNAME, then you must specify it using the dbalias: attribute.

The remote SRS web server must be specified using the url: attribute.

Database definitions should define this as methodentry or methodquery to avoid returning the entire database. Failure to do so could lead to a request to return the entire database. Although an SRS web server can cope with this, EMBOSS would then have to keep the entire web page in memory before stripping out HTML tags in order to read the first entry.

For example:

```
DB mydb [
    type:          "N"
    format:        "embl"
    methodquery:   "srswww"
    dbalias:       "embl"
    fields:        "sv des org key"
    url:           "http://srs.redbrick.ac.uk/srsbin/cgi-bin/wgetz"

# define 'all' access method
    methodall:     "direct"
    directory:     "/data/embl"
    file:          "*.seq"
]
```

Various database definitions for remote retrieval of sequences over the web via SRS are shown below:

```
DB embl [
type:     "N"
method:   "srswww"
format:   "embl"
release:  "EBI"
url:      "http://srs.ebi.ac.uk/srsbin/cgi-bin/wgetz"
comment:  "EMBL from the EBI" ]
```

```
DB em [
type:    "N"
method:  "srswww"
format:  "embl"
release: "EBI"
url:     "http://srs.ebi.ac.uk/srsbin/cgi-bin/wgetz"
dbalias: "EMBL"
comment: "EMBL from the EBI" ]

DB uniprot [
type:    "P"
method:  "srswww"
format:  "swiss"
release: "EBI"
url:     "http://srs.ebi.ac.uk/srsbin/cgi-bin/wgetz"
comment: "UNIPROT from the EBI" ]

DB uni [
type:    "P"
method:  "srswww"
format:  "swiss"
release: "EBI"
url:     "http://srs.ebi.ac.uk/srsbin/cgi-bin/wgetz/"
dbalias: "UNIPROT"
comment: "UNIPROT from the EBI" ]
```

4.3.2.6   BLAST

> **Note**
>
> Currently **dbiblast** can't use the new (format 4) style of **BLAST** indexes. You must create the old (format 3) style of **BLAST** indexes by adding **-A F** to the **formatdb** command line.

Uses an EMBLCD index from the program **dbiblast** to access databases in **BLAST** format. The **BLAST** database can be DNA or protein, produced by **formatdb**, **pressdb** or **setdb**, with or without the original FASTA format file. This can cope with all levels of access. Queries use the index files, reading all entries uses the list of files in the division.lkp file and opens each in turn.

Supports queries by

    id
    acc
    sv
    des

(*Not* by key and org as there is no way to find these in the **BLAST** database description line).

The directory containing the **BLAST** index files (`*.nin`, `*.pin`, `*.nhr`, `*.nsq`, `*.phr`, pin, psq, etc.) and the index files produced by **dbiblast** must be specified using the `directory:` attribute. If the **dbiblast** indexes are in a directory other than the one containing the **BLAST** index files, then the **dbiblast** index directory can be explicitly set using the `indexdirectory:` attribute. The available fields should be specified using the `fields:` attribute if more than just the default ID name and accession number fields have been indexed.

A wildcard search for unique fields (`id` or `sv`), or any search for `acc`, `des`, `org` or `key` is type `query` and returns a list of entries. A search for a single `id` or `sv` is of type `entry` and will find the first match in the index and assume no other matches. The ID has to be unique in an EMBLCD database.

For example:

```
DB mydb [
  type: N
  format: embl
  method: blast
  fields: "sv des"
  directory: /data/embl
]
```

4.3.2.7   EMBOSSGCG

Uses a b+tree index from the program **dbxgcg** to access a database reformatted for GCG 8, 9 or 10 by GCG programs such as **embltogcg**. As only the `.ref` and `.seq` files are used, any 'GCG' distribution of the databases can be used with **dbxgcg** without the need to create GCG-specific index files. This can cope with all levels of access. Queries use the index files. Reading all entries uses the list of files in the `[database].ent` index files and open each in turn.

Supports queries by

```
id
acc
sv
key
org
des
```

The directory containing the sequence files and indexes to be read must be specified using the `directory:` attribute. If the indexes are in a directory other than the one containing the sequence files, then the index directory can be explicitly set using the `indexdirectory:` attribute. The available fields should be specified using the `fields:` attribute if more than just the default ID name and accession number fields have been indexed.

As the b+tree indexes allow duplicate keys then all queries may return a list of entries, i.e. type `query` is used throughout.

For example:

```
DB mydb [
  type: "N"
  format: "embl"
  method: "embossgcg"
  fields: "sv des org key"
  directory: "/data/gcg/gcgembl"
]
```

You can use the file: and exclude: attributes to create file-specific subsets from a single index.

4.3.2.8 GCG

Uses an EMBLCD index from the program **dbigcg** to access a database reformatted for GCG 8, 9 or 10 by GCG programs such as **embltogcg**. As only the .ref and .seq files are used, any 'GCG' distribution of the databases can be used with **dbigcg** without the need to create GCG-specific index files. This can cope with all levels of access. Queries use the index files. Reading all entries uses the list of files in the division.lkp index files and open each in turn.

Supports queries by

id

acc

sv

key

org

des

The directory containing the sequence files and indices to be read must be specified using the directory: attribute. If the indices are in a directory other than the one containing the sequence files, then the index directory can be explicitly set using the indexdirectory: attribute. The available fields should be specified using the fields: attribute if more than just the default ID name and accession number fields have been indexed.

A wildcard search for unique fields (id or sv), or any search for acc, des, org or key is type query and returns a list of entries. A search for a single id or sv is of type entry and will find the first match in the index and assume no other matches. The ID has to be unique in an EMBLCD database.

For example:

```
DB mydb [
  type: "N"
  format: "embl"
  method: "gcg"
  fields: "sv des org key"
  directory: "/data/gcg/gcgembl"
]
```

You can use the file: and exclude: attributes to create file-specific subsets from a single index.

4.3.2.9 DIRECT

Opens database flatfile(s) and returns each entry sequentially.

This method assumes there is no indexing done on the data, so it can only process all entries – you should explicitly set up other methods for entry and query access to the same database if these are required. It is possible to access a database with the direct: method and an ID or field in the USA, but EMBOSS will read the entire database to look for matching entries if no other method is specified.

The directory containing the sequence files to be read must be specified using the directory: attribute. The files to be read must be specified using the file: attribute. You may use the exclude: attribute to exclude some selected files from consideration.

EMBL can be defined as *.dat to avoid adding the explicit filenames, e.g. est18, hum3, htg2.

If the file format supports additional fields, they can be included in the definition as fields: to allow their use in USAs.

For example:

```
DB mydb [
  type: N
  format: embl
  methodall: direct
  directory: /data/embl
  file: *.dat
]
```

4.3.2.10 URL

Uses any other web server (for example the EBI's emblfetch or swissfetch queries) to return an entry.

The remote web server's URL must be specified using the url: attribute. This URL is expected to contain one or more instances of the character pair '%s' – each of these pairs is replaced by the value of the ID name when this database is accessed. Any HTML formatting will be stripped from the resulting web page. For example:

```
DB mydb [
  type: N
  format: embl
  methodentry: url
  url: "http://server.commercial.com/cgi-bin/getseq?%s&format=embl"
]
```

4.3.2.11 APP

Runs an external application or a simple script which returns one/more/all entries. The application can be in the user's path or have an explicit path provided. EXTERNAL is the same thing as APP, but it is obsolete and its use is discouraged.

The database definition must have app: defined to specify the application command.

The database and entry name will be appended to the application command as

```
application dbname:entry
```

Both ID and accession number can be used to specify the entry. Alternatively, if the app: attribute value contains the character pair '%s', it is replaced by the value of the ID name or accession number when this database is accessed. You can also use GCG's **typedata** as an external application, to save reindexing a GCG database.

This could be a good way to search a set of databases, for example to get the first entry from *SwissNew, SwissProt, TrEmbl* and *TrEmblNew* with the ID, accession number or PID as the entryname.

For example:

```
DB mydb [
  type: "N"
  format: "embl"
  method: "app"
  app: "/usr/local/bin/accessdb -db embl -query %s"
]
```

4.3.3   Mixed access methods

For any given method: declaration, EMBOSS will use that method for those access modes supported by the method. If you wish to specify which query level (all, query or single) should be handled by which database retrieval method then the methodsingle:, methodquery: and methodall: declarations should be used instead of method:.

For example:

```
DB mydb [
methodsingle: "app"
format:       "fasta"
app:          "customapp myproteindb"
methodall:    "direct"
dir:          $emboss_db_dir/myproteindb
file:         "myproteindb.dat"
type:         "P"
comment:      "single and all access for myproteindb"
]
```

You can mix these, for example, to use a script to query a file and direct access to read all entries. For example:

```
methodall: 'direct'
methodquery: 'app'
```

4.3.4 Database farms

Currently there is no simple way of defining several data sources that could be defined as a single, composite database. The closest you can get is to define a database that calls an application that can return sequences from any one of a set of previously defined EMBOSS databases.

A script has been developed for this task by Simon Andrews. It is shown below or you can download it as the file http://emboss.open-bio.org/downloads/databasefarm.sh.

```perl
#!/usr/bin/perl -w
#
# change the above line to match the location of perl on your system
#

use strict;

# EMBOSS farm file script
#
# Written by Simon Andrews
# simon.andrews@bbsrc.ac.uk
# Dec 2001
#
# This script allows you to set up a farm
# of EMBOSS databases which can be queried
# by a single instance of seqret. The
# program must be accompanied by an entry
# in emboss.default which looks like this:
#
# DB name_of_database [
#       type: N (or P if we're dealing with proteins)
#       method: app
#       format: fasta
#       app: "/path/to/this/emboss_farm.script"
#       comment: "Whatever text you'd like to see in showdb"
# ]
#

# First we need to set a few preferences
#
# What is the full path to seqret?
# If you are sure that seqret will always
# be somewhere in your path, then you can
# just leave this as 'seqret'.

my $seqret_path = 'seqret';

# Now we need to know the names of the
# databases you'd like included in the
# search. These must be dabases which
# have already been indexed, and installed
# correctly into emboss.default. Simply
# enter the database names between the
# brackets, separated by spaces.

my @databases = qw(dbase1 dbase2 dbase3);

##### End of bits which need to be edited #########
my ($reference) = @ARGV;
if ($reference =~ /:(.+)$/) {

$reference = $1;
}
```

```
else {
   die "\n*** FARM ERROR *** Couldn't get accession after : from
$reference\n\n";
}

foreach my $database (@databases) {

  my $sequence = '$seqret_path $database:$reference fasta::stdout 2>/dev/null';

  if ($sequence) {
     print $sequence;
     exit;
  }

}
warn "\n*** FARM ERROR *** Couldn't find $reference in any of '@databases'\n\n";
```

To use this simply copy and paste the text of the script to a file on your system, then make sure that this file is readable and executable by everyone (**chmod 755 *filename***). The comments in the script tell you what changes you need to make to the script itself, and the format of the entry you need to create in emboss.default.

It will work with **seqret** (and will output any format you like), and can also be used as part of a USA for any of the standard EMBOSS programs. The script requires a UNIX-like operating system, but could trivially be adapted to run under Win32.

4.4   Miscellaneous database integration

EMBOSS can be integrated with several common non-sequence biological databases. These are described in this section.

4.4.1   REBASE

REBASE is the restriction enzyme database maintained by New England Biolabs. It is needed for programs such as **remap** and **restrict**. The latest version of Rebase can be obtained by anonymous FTP (ftp.neb.com/pub/rebase/). EMBOSS needs the withrefm and proto files. The data is extracted for EMBOSS with the program **rebaseextract**:

```
% mkdir /site/prog/emboss/data/REBASE
% rebaseextract
Extract data from REBASE
REBASE database withrefm file: /data/rebase/withrefm.208
REBASE database proto file: /data/rebase/proto.208
```

REBASE is now installed and ready to use.

4.4.2   TRANSFAC

TRANSFAC is the transcription factor binding site database. It is available by anonymous FTP (ftp.ebi.ac.uk/pub/databases/transfac/). Unpacking the distribution reveals a file called site.dat. This is the one EMBOSS needs. Run **tfextract** to extract the data from *TRANSFAC*:

```
% tfextract
Extract data from TRANSFAC
Full pathname of transfac SITE.DAT: /databases/transfac/site.dat
```

tfscan can now access the *TRANSFAC* database.

4.4.3 PROSITE

PROSITE is a database of regular expressions that match potentially diagnostic regions for structural/functional classification of proteins. EMBOSS needs this database for the patmatmotifs program. *PROSITE* can be obtained via anonymous FTP from the EMBL-EBI. Download the prosite.dat and prosite.doc files to the same directory. Then run **prosextract** to build the EMBOSS *PROSITE* database specifying the download directory:

```
% prosextract
Builds the PROSITE motif database for patmatmotifs to search
Enter name of prosite directory: /data/prosite
```

PROSITE is now integrated into your EMBOSS installation.

4.4.4 PRINTS

PRINTS is a database of diagnostic patterns of blocks of sequence homology in protein families. The *PRINTS* database can be searched using the EMBOSS program **pscan**. *PRINTS* can be obtained via anonymous FTP from the EMBL-EBI. The database is made available as compressed files which should be uncompressed using **gzip** before integrating them into EMBOSS. *PRINTS* is integrated with EMBOSS using the program **printsextract**:

```
% printsextract
Extract data from PRINTS
Input file: /data/prints/prints28_0.dat
```

The *PRINTS* database is now integrated with EMBOSS.

4.4.5 AAINDEX

An amino acid index is a set of 20 numerical values representing any of the different physicochemical and biological properties of amino acids. The **AAINDEX1** section of the *Amino Acid Index Database* is a collection of published indices together with the result of cluster analysis using the correlation coefficient as the distance between two indices. This section currently contains 437 indices in release 4.0 of the database.

The EMBOSS programs **pepwindow** and **pepwindowall** plot hydrophobicity using the data from an **AAINDEX** entry. If **AAINDEX** is installed these programs can plot the other amino acid properties.

AAINDEX can be obtained via anonymous FTP (http://www.genome.jp/aaindex/ and is integrated with EMBOSS using the program **aaindexextract**:

```
% aaindexextract
Extract data from AAINDEX
Full pathname of file aaindex1: /data/aaindex/aaindex1
```

The *AAINDEX* database is now integrated with EMBOSS.

4.4.6   CUTG

The *CUTG* database contains a series of codon usage tables calculated from *GenBank*. *CUTG* can be obtained via anonymous FTP from the EMBL-EBI server. *CUTG* is integrated with EMBOSS using the program **cutgextract** which writes files to the CODONS data directory:

```
% cutgextract
Extract data from CUTG
CUTG directory [.]: /data/cutg/
```

The *CUTG* database is now integrated with EMBOSS.

4.4.7   JASPAR

Download and unzip the Archive.zip file and then run **jaspextract** specifying the FlatFileDir directory.

```
% jaspextract
Extract data from JASPAR
JASPAR database directory [.]: /data/jaspar/all_data/FlatFileDir
```

See http://jaspar.genereg.net/html/DOWNLOAD/.

4.4.8   Miscellaneous data files

Other data files should be kept in the data directory under the main EMBOSS installation. Personal (user) data files can be kept in:

- The current working directory
- A sub-directory .embossdata of the current directory
- Their home directory
- A sub-directory .embossdata of their home directory

EMBOSS will search these locations in this order and will stop as soon as it finds a matching file. If the personal directories do not contain the desired file, EMBOSS will search the system-wide data directory (/share/EMBOSS/data/).

Apparently inexplicable errors when running EMBOSS programs may be caused by the system not using the data files one expects. The search path can be displayed in search order using the command **embossdata**.

For more information on EMBOSS data files, see the *EMBOSS User's Guide*.

4.5 Database indexing

4.5.1 Introduction

To gain experience in database indexing under EMBOSS, you can practice with the example databases included in the EMBOSS distribution. These include:

- `test/data`
- `test/embl`
- `test/pir`
- `test/swiss`
- `test/swnew`
- `test/wormpep`

You can reindex these files using the **dbx*** or the **dbi*** programs.
The **dbx*** applications are preferred.

4.5.2 Resource definitions, cachesize and pagesize

The **dbx*** programs require two variables to be set in the emboss.default file and at least one *Resource Definition* to be present. In contrast the **dbi*** programs do not require these definitions.

For example:

```
SET PAGESIZE 2048
SET CACHESIZE 200

RES embl
[
   type: Index
   idlen: 15
   acclen: 15
   svlen: 15
   keylen: 25
   deslen: 25
   orglen: 25
]
```

The **dbx*** applications buffer disc pages in order to improve performance. The PAGESIZE should usually be set to the size, in bytes, that your operating system uses to buffer disk pages, though the value is not critical. The CACHESIZE should be set to the number of such pages that you wish to be cached. The values of 2048 and 200 given above are good general-purpose ones. We recommend a CACHESIZE greater than 100.

You should have at least one Resource Definition (RES entry) in your emboss.default file, though we recommend having one per database you wish to index. The **dbx*** programs will ask for the name of a RES entry when they run. The definitions have a compulsory type: Index attribute followed by length attributes for each of the fields that can be indexed. These lengths represent the maximum length of the field before potential truncation occurs. Truncation of ID keys is usually to be avoided as it can lead to duplicate IDs being indexed.

It is appropriate to set the `idlen`, `acclen` and `svlen` attributes a little larger than the maximum size field you expect in the source file. Values for `keylen`, `deslen` and `orglen` are more a matter of preference.

4.5.3 Indexing and configuration

4.5.3.1 Flatfile databases

Flatfile databases are plain text files in a defined format such as those released by *EMBL*, *GenBank*, etc. The EMBOSS program **dbxflat** is used to generate EMBOSS indexes that can be used for all types of database access. The **dbiflat** application can also be used but cannot cope with large source database files (greater than 2 Gb) or duplicate IDs or accession numbers.

dbxflat (and the EMBOSS access method) requires the databases to be uncompressed. The examples given here will not probe the deeper secrets of **dbxflat** (for which the reader is referred to the application documentation, or failing that the source code) but will show a typical installation for a common database.

We assume that EMBOSS has been installed and works. This can be tested with the command:

```
wossname -auto
```

which should list all the programs available.

In this example you will index and configure the *EMBL* database for use with EMBOSS. First download and unpack the *EMBL* database. This will require a considerable amount of disk space. If you do not have sufficient space available then just download a subset of the database. Use **cd** to move the directory in which you have unpacked *EMBL*. This should look something like this when you run **ls**:

```
% ls
.
rel_est_fun_01_r98.dat
rel_est_fun_02_r98.dat
rel_est_fun_03_r98.dat
.
Output truncated
.
wgs_cabc_pro.dat
wgs_cabd_mam.dat
wgs_cabe_fun.dat
```

Run **dbxflat** to create the EMBOSS indices. This assumes you have set up a RES definition and cache and page sizes as described above.

```
% dbxflat

Index a flat file database using b+tree indices
Basename for index files: embl
Resource name: embl
```

```
      EMBL : EMBL
     SWISS : Swiss-Prot, SpTrEMBL, TrEMBLnew
        GB : Genbank, DDBJ
    REFSEQ : Refseq
Entry format [SWISS] : EMBL
Wildcard database filename: *.dat
Database directory [.]: .
        id : ID
       acc : Accession number
        sv : Sequence Version and GI
       des : Description
       key : Keywords
       org : Taxonomy
Index fields [id,acc] : id,acc
General log output file [outfile.dbxflat] : embllog.dbxflat
```

dbxflat should happily chug away for some considerable time (depending on the speed of your machine) and will generate (eventually) the following index files:

```
% ls
embl.ent
embl.xid
embl.xac
embl.pxid
embl.pxac
embllog.dbxflat
```

Now create an entry in the EMBOSS configuration files to access the database. It is probably a good idea to try new database definition in your local configuration file first. Put the following entry in your .embossrc:

```
DB embl
[
   type:      "Nucleotide"
   method:    "emboss"
   format:    "embl"
   directory: "$emboss_db_dir/embl"
   filename:  "*.dat"
   release:   "98.0"
   comment:   "EMBL release 98.0"
]
```

You will have needed to predefine $emboss_db_dir somewhere in your emboss. default or .embossrc using a directive such as:

```
set emboss_db_dir /path_to_databases
```

Save .embossrc and try running **showdb**. You should see a line that looks like:

```
% showdb
.. output deleted
embl     N   OK OK OK    EMBL release 63.0
.. output deleted
```

4.5.3.1.1 Fine-tuning the installation

It can be a good idea to set up subsections of the database so that end-users can search just the regions they wish to search. This section applies to all access methods (Section 4.3, 'Database access methods') that use EMBOSS style indexes and to others as well (e.g. EMBLCD).

Files can be included with the declaration:

```
filename:
```

or excluded with the declaration

```
exclude:
```

In order to just take the expressed sequence tags (EST) files in our *EMBL* database try the following:

```
DB emblest
[
   type:        "Nucleotide"
   method:      "emboss"
   format:      "embl"
   directory:   "$emboss_db_dir/embl"
   filename:    "rel_est*.dat"
   release:     "98.0"
   comment:     "EMBL release 98.0"
]
```

Files can also be given as a space-separated list enclosed in quotes. For example, to set up a database of all mammalian sequences (except genomes) try the following:

```
DB emblallmam
[
   type:        "Nucleotide"
   method:      "emboss"
   format:      "embl"
   directory:   "$emboss_db_dir/embl"
   filename:    "rel_std_rod*.dat rel_std_mus*.dat rel_std_hum*.dat rel_std_mam*.dat"
   release:     "98.0"
   comment:     "EMBL release 98.0"
]
```

As you can see from these two examples, the `filename:` tag takes a space delimited list of filenames enclosed in quotes that can contain normal wildcard (?*) characters. It can be quite tedious to set up a long list of sequences to search. In many cases you can use the `exclude:` tag to make things easier:

```
DB emblnoest
[
   type:      "Nucleotide"
   method:    "emboss"
   format:    "embl"
   directory: "$emboss_db_dir/embl"
   filename:  "*.dat"
   exclude:   "rel_est*.dat"
   release:   "98.0"
   comment:   "EMBL release 98.0"
]
```

This configures the *emblnoest* database to contain all of *EMBL* except the ESTs.

4.5.3.2   GCG format databases

EMBOSS can access GCG formatted databases, thus avoiding having multiple copies of the same databases in different formats for those who still use GCG alongside the flatfiles. EMBOSS creates b+tree indices for the GCG format databases using the program **dbxgcg**. This runs in much the same way as **dbxflat**. You will need the GCG format `.seq` and `.ref` files in order to create an EMBOSS indexed database.

Move to the GCG database directory containing your data and run **dbxgcg**:

```
% dbxgcg
Index a GCG formatted database
Basename for index files: emblgcg
Resource name: embl
EMBL : EMBL
SWISS : Swiss-Prot, SpTrEMBL, TrEMBLnew
GENBANK : Genbank, DDBJ
PIR : NBRF
Entry format [SWISS]: embl
Database directory [.]:
Wildcard database filename [*.seq] : *.seq
Wildcard database filename [*.seq] :
        id : ID
       acc : Accession number
        sv : Sequence Version and GI
       des : Description
       key : Keywords
       org : Taxonomy
Index fields [id, acc] :
General log output file [outfile.dbxgcg] : emblgcglog.dbxgxg
```

When **dbxgcg** prompts for the entry format:

```
Entry format [EMBL] :
```

you should enter the original database format before you ran **embltogcg** or similar to generate the GCG databases. The program will run for a while and will then generate the EMBOSS index files for the GCG format database.

The following entry should be put in your .embossrc file:

```
DB gcgembl
[
    type:       "Nucleotide"
    method:     "embossgcg"
    format:     "embl"
    directory:  "$emboss_db_dir/embl"
    filename:   "*.dat"
    release:    "98.0"
    comment:    "EMBL release 98.0"
]
```

showdb should show your newly configured database.

You can configure subsets of the databases in the same way as for the original format databases, as described above. One difference to **dbxflat** indexing is that both the .seq and .header files are listed in the [database].ent file. The filename: and exclude: directives should therefore be of the form:

```
exclude:
*/rel_est*
```

instead of just:

```
*/rel_est*.seq
```

4.5.3.3   BLAST databases

BLAST format databases are generated for efficient homology searching using the **BLAST** programs. It can be convenient to avoid redundant copies of databases so EMBOSS provides a mechanism for accessing these databases.

BLAST format databases are those generated using the tools distributed with **NCBI-BLAST** or with **WU-BLAST**.

For indexing of one **BLAST** database, move to the directory containing your **BLAST** format databases and run **dbiblast**:

```
% dbiblast
Index a BLAST database
Database name: blastsw
Database directory [.]:
database base filename [blastsw]:
Release number [0.0]:
Index date [00/00/00]:
         N : nucleic
         P : protein
         ? : unknown
```

```
Sequence type [unknown] : p
        1 : wublast and setdb/pressdb
        2 : formatdb
        0 : unknown
Blast index version [unknown] : 2
```

The program will run for a while and will then generate the EMBLCD index files for the **BLAST** format database.

The following entry (or one like it that is more appropriate to your particular installation) should be put in your .embossrc file:

```
DB blastsw
[
   type:       "Protein"
   method:     "blast"
   format:     "ncbi"
   directory:  "$emboss_db_dir/blastsw"
   filename:   "blastsw"
   release:    "38.9"
   comment:    "BLAST format Swissprot"
]
```

showdb should show your newly configured database.

Because of the way **BLAST** works, many sites may group their **BLAST** databases in the same directory. You can index these *in situ* with **dbiblast** but this may require some extra steps if your databases are not of the same type; generation of subsequent index files will overwrite those that already exist. To avoid overwriting of index files you can index many databases with one set of index files, or you can use the **-indexdir** options to place the indexes in a different directory.

There are two requirements for indexing several databases together in one index. The first is that the databases are the same type (protein/nucleic acid) and generated with the same tool (**pressdb** or **formatdb**); the second is that all the ID and accession numbers in the combined databases are unique.

Run **dbiblast** as before but specify all the databases you wish to be included when prompted for the database filename:

```
% dbiblast
Index a BLAST database
Database name: alldbs
Database directory [.] :
database base filename [alldbs] : dbone dbtwo dbthree dbfour
Release number [0.0] :
Index date [00/00/00] :
        N : nucleic
        P : protein
        ? : unknown
Sequence type [unknown] : p
        1 : wublast and setdb/pressdb
        2 : formatdb
        0 : unknown
Blast index version [unknown] : 2
```

These can then be configured by using the `filename:` and `exclude:` tags as appropriate.

When you have databases of different types, generated with different programs or where the ID/accession numbers are duplicated between databases the preferred strategy is probably to keep the source data for the individual databases in separate directories and index them there.

Alternatively you can place the index files in a separate directory. This requires that you run **dbiblast** with the **-indexdirectory** and set the `indexdirectory:` tag in the database configuration to point to the correct database.

The example below illustrates database configuration using the **indexdir** options:

```
% dbiblast -indexdir /databases/indices/mydb
Index a BLAST database
Database name: mydb
Database directory [.]:
database base filename [mydb]:
Release number [0.0]:
Index date [00/00/00]:
         N : nucleic
         P : protein
         ? : unknown
Sequence type [unknown]: p
         1 : wublast and setdb/pressdb
         2 : formatdb
         0 : unknown
Blast index version [unknown]: 2
```

The corresponding entry in .embossrc or emboss.default would look like:

```
DB mydb
[
   type:              "Protein"
   method:            "blast"
   format:            "ncbi"
   directory:         "$emboss_db_dir/blastsw"
   indexdirectory:    "/databases/indices/mydb"
   filename:          "mydb"
   release:           "1.0"
   comment:           "My BLAST DB with an index in a different directory"
]
```

Again, multiple indexes cannot coexist in the same directory so care should be taken when using the **-indexdir** option that an existing database index is not overwritten.

4.5.3.4 FASTA databases

The FASTA specifications just define the sequence file as a header line that begins with > and subsequent lines contain the sequence. The header line can be present in a seemingly infinite number of formats, several of which can be processed by EMBOSS. EMBOSS attempts to determine the accession number and/or ID for each sequence. For indexing purposes there is no semantic difference between an accession number and an ID. In the real world, accession numbers should be immutable, i.e. they do not change with subsequent releases of the database, but IDs may change.

One of the programs that can be used to process FASTA format databases is **dbxfasta**. It can recognise the following header line formats, specified on the command line:
simple.

```
>id...
```

idacc.

```
>id accno...
```

gcgid.

```
>db:id...
```

gcgidacc.

```
>db:id acc...
```

dbid.

```
>db id...
```

ncbi.

```
>...[|accno]|id...
```

Other header formats will not be recognised by **dbxfasta** and will cause indexing and/or database lookup to fail. If you have a header format that **dbxfasta** cannot yet handle you have two options:

1. (The preferred option) Get a C programmer to modify the source code for **dbxfasta** and recompile. If you are a community-spirited person you will also contribute these changes to the main EMBOSS source tree (email emboss-dev@emboss.open-bio.org for more information on contributing changes to the EMBOSS source code and/or read the EMBOSS developers documentation).

2. (The quick hack) Write a custom script (using e.g. BioPerl http://www.bioperl.org) to access your database and use method: external to configure it. This is less desirable as you may be limited in the access modes you can use.

To index a FASTA format database, run **dbxfasta**:

```
% dbxfasta
Index a fasta file database using b+tree indices
Basename for index files: mydb
Resource name: myresdef
```

```
     simple : >ID
      idacc : >ID ACC or >ID (ACC)
      gcgid : >db:ID
   gcgidacc : >db:ID ACC
        dbid : >db ID
        ncbi : | formats
ID line format [idacc] : idacc
Database directory [.] :
Wildcard database filename [*.dat] : mydb.fasta
        id : ID
       acc : Accession number
        sv : Sequence Version and GI
       des : Description
Index fields [id, acc] : id, acc
General log output file [outfile.dbxfasta] : mydb.dbxfasta
```

dbxfasta will run for a while and will produce the index files. You can use the same
`-indexdir` options as for **dbxflat, dbxgcg** and **dbiblast** to place the indexes in a different
directory.

Place (e.g.) the following entry in your `.embossrc`:

```
DB mydb
[
    type:        "Protein"
    method:      "emboss"
    format:      "fasta"
    directory:   "$emboss_db_dir/mydb"
    filename:    "mydb.fasta"
    comment:     "My database"
]
```

`format:` should be `dbid`, `ncbi` or `fasta` (the latter for every format except `dbid` or `ncbi`).
The same `filename:` and `include:` tags can be used as for the other database indexing
programs.

4.5.3.5   Other databases

Many institutions may have local databases set up in their own Laboratory Information
Management System. EMBOSS provides a simple mechanism for interfacing with such
systems.

As long as a program is available that can be called non-interactively and returns the
specified sequence on standard output, EMBOSS can interface with it. Use `method:` app
and `app: program` command. The ID given in the USA will be appended to the command
used to run the program. It is often best to specify the methods available using the method
subsets, `methodall:`, `methodquery:` and `methodsingle:` rather than using the
generic `method:` tag.

4.5.4   Configuring EMBOSS to use SRS for database lookup

SRS is a powerful database querying system that can cross reference different databases,
launch applications, etc. SRS can be run either through a web interface (see the description of
the SRSWWW method above for an example) or via the command line program **getz**.

Indexing and configuring databases for SRS is not described here, just how to connect to preconfigured and indexed SRS databases. If **getz** is already within the scope of your PATH environment variable then insert the following (or similar) into your .embossrc file:

```
DB emblgetz
  [
  type: N
  method: srs
  release: "98"
  format: embl
  comment: 'EMBL using getz'
  dbalias: embl
  app: getz
]
```

This will provide access to the SRS database *embl* as emblgetz:acc. If the SRS database has a different name from the DBNAME (as is the case here) then the dbalias: tag should be used to access the correct SRS database.

This configuration can be extremely slow for the all access mode. It is probably a better idea to set up the database as follows:

```
DB emblgetz
[
  type:          "Nucleotide"
  methodquery:   "srs"
  release:       "63"
  format:        "embl"
  comment:       "EMBL using getz"
  dbalias:       "embl"
  app:           "getz"
  methodall:     "direct"
  filename:      "*.dat"
  directory:     "$emboss_db_dir/embl"
]
```

This will use method: srs for the query access mode but will use method: direct for the all access mode, thus speeding up reading of the whole database.

The **SRSFASTA** access method is identical to the normal **SRS** method except that it returns the sequence in FASTA format and so does not need a format: tag.

4.5.5   Size of the dbx* indexes

You might notice that the index files produced by the **dbx*** applications can be very large. This is normal and is a consequence of three things. First, a tree structure is used, second the tree isn't tightly packed and third 64-bit pointers are used throughout. The first will allow on-the-fly updating of the index, the second is for speed of construction/updating and the third is obvious. Another consideration is that, in some cases, the indexes are trees-of-trees to allow duplicate codes to be indexed (e.g. keywords).

Appendix A   Resources

A.1   Software distributions that include EMBOSS

A.1.1   RPM Package Manager files

The RPM Package Manager (RPM) is a command line driven package management system for installing, uninstalling, verifying, querying and updating software packages. Each software package consists of an archive of files along with package information such as version, description etc. For more information about RPM see http://www.rpm.org/.

Ryan Golhar (Informatics Institute, University of Medicine and Dentistry of New Jersey) has built RPMs for EMBOSS 5.0.0 and related applications. They are available from

```
http://informatics.umdnj.edu/BioRPMs/
```

or from the EMBOSS FTP site:

```
ftp://emboss.open-bio.org/pub/EMBOSS/contrib/
```

Caution

Files under contrib/ are contributed to the EMBOSS project but are not part of the EMBOSS or EMBASSY distributions. Owing to the limited availability of test platforms, these contributions might not have been tested fully or even at all by the core EMBOSS developers. In case of difficulty, please contact the original author, in this case, *Ryan Golhar*.

A.1.2   Ports and packages

The lists below are by no means comprehensive. If you know of other relevant links please email emboss-bug@emboss.open-bio.org with details.

Caution

Sites who have packaged EMBOSS may not have kept that package up to date.

Live DVD with EMBOSS/**Jemboss**	Live DVD with EMBOSS/**Jemboss** from NERC Environmental Bioinformatics Centre See http://nebc.nox.ac.uk/tools/bio-linux
Another live CD with EMBOSS	Live DVD with EMBOSS/**Jemboss** from dnalinux.com See http://www.dnalinux.com/
Bio::Emboss Perl module version 1.0	This module is an interface to the libraries of the EMBOSS package. With this module you can access EMBOSS databases and use EMBOSS functions to manipulate your data. You can write Perl programs that look like any other EMBOSS program (command line, web interface). See http://search.cpan.org/dist/Bio-Emboss/
Bio::Tools::Run:: EMBOSSApplication	The EMBOSS factory class encapsulates access to EMBOSS programs. A factory object allows creation of only known applications. See http://doc.bioperl.org/bioperl-run/Bio/Factory/EMBOSS.html.
FreeBSD port	FreeBSD port of EMBOSS See http://www.freebsd.org/cgi/cvsweb.cgi/ports/biology/emboss/.

Appendix B   EMBOSS Frequently Asked Questions

EMBOSS maintains a list of frequently asked questions (FAQs) with answers in the file FAQ in the EMBOSS distribution, e.g.

```
auser/emboss/emboss/FAQ
```

That file is reproduced here. The FAQ is organised as follows:

Q & A B.1.1, 'General'	General EMBOSS issues, licensing, availability, etc. or any areas not covered by other FAQs.
Q & A B.1.2, 'Getting started'	Hardware and software requirements, downloading, etc.
Q & A B.1.3, 'Administration (installation and compilation)'	Installation and compilation.
Q & A B.1.4, 'Databases'	Databases configuration.
Q & A B.1.5, 'Administration (other)'	Post-installation setup and other administration issues not covered elsewhere.
Q & A B.1.6, 'Features'	EMBOSS key features for biologist end-users, how to request features, etc.
Q & A B.1.7, 'Sequence files and formats'	Hardware and software requirements, downloading, installing and compiling, post-compilation setup, etc.
Q & A B.1.8, 'Help and support'	EMBOSS support, mailing lists, reporting bugs, requesting features, training courses, etc.
Q & A B.1.9, 'Documentation'	EMBOSS documentation.
Q & A B.1.10, 'Applications'	Using the applications.
Q & A B.1.11, 'EMBASSY'	EMBASSY packages and applications.
Q & A B.1.12, 'ACD'	ACD syntax and files.
Q & A B.1.13, 'Interfaces'	Use of EMBOSS interfaces.
Q & A B.1.14, 'Graphics'	Graphics.

Q & A B.1.15, 'Internals' EMBOSS internals and core features for software developers.

Q & A B.1.16, 'Software development' EMBOSS software development, programming libraries, etc.

B.1.1 General

Q: Citation: Is there a reference I can cite for EMBOSS?

A: Rice P., Longden I. and Bleasby A. EMBOSS: The European Molecular Biology Open Software Suite. *Trends in Genetics* 2000 **16**(6):276–277.

B.1.2 Getting started

Q: Can I get the latest code via CVS ?

A: Yes. Here is the information you will need. Make sure you have cvs on your system. Then log into the cvs server at open-bio.org as: user cvs with password cvs.

```
cvs -d :pserver:cvs@cvs.open-bio.org:/home/repository/emboss login
```

The password is cvs. To checkout the EMBOSS source code tree, put yourself in a local directory just above where you want to see the EMBOSS directory created. For example if you wanted EMBOSS to be seen as /home/joe/src/emboss... then cd into /home/joe/src then checkout the repository by typing:

```
cvs -d :pserver:cvs@cvs.open-bio.org:/home/repository/emboss checkout
emboss
```

Or if you want to update a previously checked-out source code tree:

```
cvs -d :pserver:cvs@cvs.open-bio.org:/home/repository/emboss update
```

You can logout from the CVS server with:

```
cvs -d :pserver:cvs@cvs.open-bio.org:/home/repository/emboss logout
```

This is a read-only server.

B.1.3 Administration (installation and compilation)

Q: How do I compile EMBOSS?

A: If this is the first time trying to compile all you need to do is:

```
./configure
make
```

The above will produce the EMBOSS programs in the emboss sub-directory and you can set your PATH variable to point there. This method is suitable for EMBOSS developers. For system-wide installations we recommend installing the EMBOSS programs into a different directory from the source code (e.g. in the directory tree /usr/local/emboss). To do this type (e.g.):

```
./configure --prefix=/usr/local/emboss
make [to make sure there are no errors]
```

then

```
make install [if there are no errors]
```

You should then add (e.g.) /usr/local/emboss/bin to your PATH variable:

```
set path=(/usr/local/emboss/bin $path) [csh/tcsh shells]
export PATH="$PATH:/usr/local/emboss/bin" [sh/bash shells]
```

If you wish to recompile the code at any stage then you should type **make clean** and then configure and make source again as appropriate.

Q: I have a Linux system and compilation ends prematurely saying that it can't find the l**X11** libraries ... but I know I have **X11** installed.

A: This should not happen with versions of EMBOSS later than v6.1.0 as the configuration will inform you of missing files and terminate at that stage. If you have a version of EMBOSS less than or equal to v6.1.0 then the problem is that you probably have the **X11** server installed but you haven't installed the **X11** development files. For example, on RPM distributions, you need to install:

```
xorg-x11-proto-devel (xorg-based X11 distros) or
XFree86-devel (XFree86-based X11 distros)
```

After installing those system files you will need to:

```
make clean
```

and re-perform the **./configure** command from the top EMBOSS source directory.

Q: I'm trying to compile EMBOSS with PNG support.

A: Your system will need to have **zlib** (http://www.zlib.net: current version is 1.2.3), **libpng** (http://www.libpng.org: current version is 1.2.35) and **gd** (http://www.libgd.org: current version is 2.0.35). The version of **gd** must be $\geqslant 2.0.28$ if it is to be used with EMBOSS.

Note that the above packages often come with your operating system distribution but may not be installed by default, i.e. they are optional packages which you can install at a later date. Also note that, as well as having the above system libraries installed, you must also have installed their associated development files (these typically have devel in the package names on Linux systems). So, on Linux RPM systems you would need to make sure that packages called names similar to **gd** and **gd-devel** are installed.

Mac OS X users can often find the above packages available on the MacPorts site (http://www.macports.org). However, for some operating systems, there may be no freeware sites where you can find pre-compiled versions of **zlib/libpng/gd** and you may have to compile one or more of them from their source distributions (URLs given above).

You can unpack the tar.gz or tar.bz2 files in any directory, and install them in a common area. By default everything (including EMBOSS) installs in /usr/local. To install, pick up the sources and then:

```
gunzip -c zlib-1.2.3.tar.gz | tar xf -
gunzip -c libpng-1.2.35.tar.gz | tar xf -
gunzip -c gd-2.0.35.tar.gz | tar xf -

cd zlib-1.2.3
./configure
make
make install
./configure --shared
make
make install
cd ..

cd libpng-1.2.35
./configure
make
make install
cd ..

cd gd-2.0.35
./configure --without-freetype --without-fontconfig
make
make install
cd ..
```

To compile with the local version your EMBOSS ./configure line should now read:

```
./configure --with-pngdriver=/usr/local
```

Q: When installing EMBOSS recently I get a load of errors due to libraries not found. The main problem is that I have an old version of **libz** but no **libgd** in my system libraries and EMBOSS looks there first to try to locate these libraries. I have the correct versions installed elsewhere. Are there any suggestions for setting the library search path or am I missing something really obvious?

A: There are the

```
--without-pngdriver
```

and

```
--with-pngdriver=dir
```

flags. Did you try them? If the libraries are in /opt/png/lib then set dir to /opt/png i.e. one level above the lib directory.

Q: How do I compile the CVS (developer's) version?

A: You will need **automake, autoconf,** (GNU) **make** and **libtool** for this. The **gcc** compiler is recommended. The host **cc** compiler should work. (Note that some non-gcc compilers may need `--include-deps` added to the automake command line.)

If you have system versions of the GNU tools that, by chance, are more recent than the GNU tools used in the EMBOSS source tree then typing `autoreconf -fiv` may be necessary before you start. Mac OS X CVS users should also use `autoreconf -fiv` after downloading the source. Then type:

```
aclocal -I m4
autoconf
automake -a # --include-deps # some non-gcc compilers
./configure # --enable-warnings is a useful developers' switch
make
```

For more information on the configurability of the build try

```
./configure --help
```

Q: Can you give an example of how to install an EMBASSY package?

A: Here is how to install **PHYLIP** given the various ways you can install the main EMBOSS package.

From the anonymous cvs code:

1. Go to the **phylip** directory

```
cd embassy/phylipnew
```

2. Make the configuration file

```
aclocal
autoconf
automake
```

3. Configure and compile

```
./configure (use same options as you used to compile emboss, especially the --
prefix))
make
make install
```

From `PHYLIP-3.68.tar.gz` available from our FTP server ftp://emboss.open-bio.org/ pub/EMBOSS/ in file `PHYLIP-3.68.tar.gz`

If you have done a full installation of EMBOSS using a 'prefix', e.g. you configured with **./configure --prefix=/usr/local/emboss** and followed this with a **make install** (highly recommended), then:

1. gunzip and untar the file anywhere:

```
gunzip PHYLIP-3.68.tar.gz
tar xf PHYLIP-3.68.tar
```

2. Go into the **phylip** directory

```
cd PHYLIP-3.68
```

3. Configure and compile

```
./configure (use same options as you used to compile emboss, especially the
--prefix))
make
make install
```

N.B. If you configured without using a prefix but did do a **make install** (or specified a prefix of /usr/local, which amounts to the same thing) then you must configure using:

```
./configure --prefix=/usr/local --enable-localforce
```

If, on the other hand, you did not do a **make install** of EMBOSS then:

1. Go to the emboss directory

```
cd EMBOSS-6.1.0
```

2. Make new directory embassy if it does not exist already.

```
mkdir embassy
```

3. Go into that directory

```
cd embassy
```

4. gunzip and untar the file

```
gunzip PHYLIP-3.68.tar.gz
tar xvf PHYLIP-3.68.tar
```

5. Go into the **phylip** directory

```
cd PHYLIP-3.68
```

6. Configure and compile

```
./configure (use same options as you used to compile emboss)
make
```

7. Set your PATH to include the full path of the src directory.

Q: I have EMBOSS installed on our development server and I'm preparing a dispatch which will send it out to about 20 remote sites. I ran the configure with the - -**prefix** option to install to a private directory. I also collected all the data files (rebase, transfac, etc.) to another directory and extracted the information from them with the relevant programs.

My intention was to simply transfer the EMBOSS install directory and the Data directory to all sites, using symlinks where necessary so that the directory paths corresponded. However, when testing this I have found a couple of problems:

1. Although the EMBOSS programs work, I can't see any of the extracted data. For instance **remap** gives the error:

```
EMBOSS An error in remap.c at line 167:
Cannot locate enzyme file. Run REBASEEXTRACT
```

This is despite the fact that I have both the EMBOSS install and the Data directories in the same place as on the development machine (which works).

2. The other major problem is that I can no longer see my databases defined in emboss. default. Again, the file exists, and is in the same place as on the development machine, but the box it is transferred to gives an empty list from **showdb**.

Does anyone know where EMBOSS stores the information about the location of these files? It can't have installed anything outside the original installation directory (wasn't installed as root), so I'm guessing that the problem stems from the program resolving symlinks at some point.

A: It is inside the binaries. EMBOSS 'knows' the location of the files because it is picked up during the configure, when you build your copy, and included in the binaries. You can see it during compilation, especially of ajnam.c (where it is used):

```
-DAJAX_FIXED_ROOT=\"/full/source/path\" -DPREFIX=\"/install/prefix/path\"
```

To copy binaries, you need to define environment variable(s) to override the compile-time definitions, unless you can make the path (e.g. /usr/local) the same for the installations at each site.

emboss.default can set environment variables too, but you need to tell EMBOSS where to find that file.

```
setenv EMBOSS_ROOT /dir/for/default/file
```

and then, in the emboss.default file you can set:

```
SET EMBOSS_ACDROOT /install/dir/share/EMBOSS/acd
```

or (this overrides it) you can use another environment variable:

```
setenv EMBOSS_ACDROOT /install/dir/share/EMBOSS/acd
```

Q: I have downloaded the EMBOSS source and installed it for use at XYZ University without any difficulty. I've had some advice on configuring the software using emboss.default, and seen some examples for allowing access to SRS indexes. That appears to be done via the program **getz**, which is not part of the EMBOSS package.

A: If you have SRS installed (so you have local SRS index files) you will have a local copy of the **getz** program, which is part of SRS.

If you do not have SRS, you can build your own index files using **dbxflat, dbxgcg** (if you have GCG), **dbiblast** (if you have **BLAST**) and **dbxfasta**. This is the usual solution for sites that have no other database indexing in use.

You can also use SRS servers remotely, to get single entries, using their URLs. No extra software is needed (EMBOSS just uses the HTTP). Of course, if you really need to build your own SRS indexes you could install it. SRS is a commercial product, but academic licences are available.

Q: I am not getting full static files even when I configure with

```
--disable-shared
```

A: This most often happens when using GNU LD. If both shared and static versions of a library exist then GNU LD will take the shared library as preferred. The root of this problem is libtool. You can, however, force complete static images by adding a definition to your "make" line:

```
make "LDFLAGS=-Wl,-static"
```

B.1.4 Databases

Q: Where are the test databases?

A: In the directory /emboss/emboss/test.

B.1.5 Administration (other)

Q: How do I use my own private data file?

A: You may wish to amend one of the standard EMBOSS data files. Some of the data files you might wish to alter are the translation table files. **transeq** has been written to only read in one of the standard translation files:

EGC.0

EGC.1

EGC.2

EGC.3

EGC.4

EGC.5

EGC.6

EGC.9

EGC.10

EGC.12

EGC.11

EGC.13

EGC.14

EGC.15

These files are the only ones that you can specify to **transeq**.

If you wish to create your own specialised translation table, then you should pick one of them to amend. For instance you may decide that you will use the file EGC.15 as you would never want to otherwise use it. Use the program **embossdata** to get a copy of this file:

```
% embossdata -fetch -filename EGC.15
Finds or fetches the data files read in by the EMBOSS programs
File '/fu/bar/emboss/data/EGC.15' has been copied successfully.
```

Edit the file EGC.15 to suit your requirements. Specify **-table 15** when you run **transeq** to use this altered file. **transeq** will then look for the file EGC.15 and will find it in your current directory before it finds the default one in the EMBOSS_DATA directory. It will therefore use your local copy.

You may get confused with many copies of changed files floating about. To check which copy of a file is being used (the default EMBOSS_DATA one or a potential local copy) use **embossdata**:

```
%embossdata -filename EGC.15
Finds or fetches the data files read in by the EMBOSS programs
# The following directories can contain EMBOSS data files.
# They are searched in the following order until the file is found.
# If the directory does not exist, then this is noted below.
# '.' is the UNIX name for your current working directory.

File ./EGC.15                            Exists
File .embossdata/EGC.15                  Does not exist
File /home/joe/EGC.15                    Does not exist
File /home/joe/.embossdata/EGC.15        Does not exist
File /usr/local/emboss/data/EGC.15       Exists
```

This shows that a copy of EGC.15 exists in your current directory and so will be used in preference to the default one in the EMBOSS-DATA directory.

Q: Can I alter the location of the ACD files?

A: EMBOSS will find the ACD files in the install directory or in the original source directory (depending on how you configured it) so there is usually no reason to change this. However, should you wish to do so then it can either be done in the `emboss.default` (or `~/.embossrc`) file:

```
env emboss_acdroot /usr/local/share/EMBOSS/acd
```

or by setting an environment variable on the command line:

```
setenv EMBOSS_ACDROOT /usr/local/share/EMBOSS/acd
```

B.1.6 Features

Q: I would like to know if EMBOSS can perform nucleotide contig assembly similar to the function that GCG **gelproject/gelview** has. And if yes, is there any size restriction on the number of base pair and the number of contigs?

A: EMBOSS does not cover the sort of contig assembly you describe. An EMBOSS wrapper for the **MIRA** package is available.

B.1.7 Sequence files and formats

Q: How do you write sequences to different files instead of writing them all to one file?

A: EMBOSS is not good at writing multiple sequences to different files. You could try using **nthseq** to pull out one sequence at a time. You should consider using the -ossingle qualifier. This writes sequences to separate FASTA files, but the file names depend on the sequence ID. This is rarely used. It is there because -ossingle is the default for GCG output format. The output filename is currently ignored when ossingle is used, and the filename depends on each individual sequence.

Q: What sequence formats are supported?

A: Tens of them(e.g.): gcg, embl, swissprot, fasta, ncbi, genbank, nbrf, codata, strider, clustal, phylip, acedb, msf, ig, staden, text, raw, asis, etc.

Q: What is the difference between "TEXT" and "RAW" formats?

A: TEXT accepts everything in the sequence file as sequence. RAW accepts only alphanumeric and whitespace characters and rejects anything else.

Q: What is 'ASIS' format?

A: The 'filename' is really the sequence. This is a quick and easy way of reading in a short fragment of sequence without having to enter it into a file. For example:

```
program -seq asis::ATGGTGAGGAGAGTTGTGATGAGA
```

Q: I have some very short protein sequences that EMBOSS thinks are nucleic acid sequences. How do I force EMBOSS to treat them as nucleic acid sequences? For example:

```
%cat seq1
   A
%cat seq2
   I

% water seq1 seq2 -stdout -auto
   Smith-Waterman local alignment.
   An error has been found: Sequence is not nucleic
```

Here, **water** automatically (and wrongly) thinks that A is adenosine instead of alanine and fails when it reads in seq2 and expects to find another nucleic acid sequence – but I is not a valid base and so it fails.

A: For many sequence formats there is no way to specify the sequence type in the file, so EMBOSS has to guess. There is a flag that can force EMBOSS programs to treat sequences as nucleic acid or protein.

```
water -help -verbose
```

shows the full list of sequence qualifiers.

  If you follow the sequence USA with -**sprotein** EMBOSS will check that it is a valid protein sequence and then treat it as such. If you need to force a sequence to be DNA, the qualifier is

```
-snucleotide
```

The qualifier must follow the sequence to apply to one sequence, or can go at the start of the command line to refer to all sequences, for example:

```
water -sprotein seq4 seq3 -stdout -auto
```

You can also use -**sprotein1** anywhere on the command line to refer to the first sequence and -**sprotein2** to refer to the second sequence. Of course, like all EMBOSS qualifiers, you can shorten them so long as they are still unique. In this case, -**sp** and -**sn** will work (or -**sp1** and -**sp2** if you need the numbers.

B.1.8 Help and support

Q: How do I report bugs?

A: Mail emboss-bug@emboss.open-bio.org

Q: How do I contact the core development team?

A: Rather than email their personal addresses, for EMBOSS matters we request that you use the address:

```
emboss-bug@emboss.open-bio.org
```

This will ensure that your email will be seen by the appropriate developer and that it will get logged with our tracking system.

Q: Are there any mailing lists about EMBOSS?

A: emboss@emboss.open-bio.org
This is an open list (anyone can join) for general announcements and discussions by users.

```
emboss-dev@emboss.open-bio.org
```

This is a closed list (subscription requests have to be approved) for discussions by developers using EMBOSS.

```
emboss-announce@emboss.open-bio.org
```

This is an open list for announcements of new releases.
You can access the archives, subscribe/unsubscribe and alter the way email is sent to you (e.g. digests) by visiting:

```
http://emboss.open-bio.org/mailman/listinfo/emboss
http://emboss.open-bio.org/mailman/listinfo/emboss-dev
http://emboss.open-bio.org/mailman/listinfo/emboss-announce
```

Q: Is there a tutorial?

A: See the EMBOSS tutorial in the *EMBOSS User's Guide*:

```
http://emboss.open-bio.org/
```

B.1.9 Documentation

Q: Where's the documentation?

A: All the documentation can be found at:

```
http://emboss.open-bio.org/
```

Q: Where's the application documentation?

A: http://emboss.open-bio.org/rel/dev/apps/ and in the EMBOSS distribution, installed under

```
share/EMBOSS/doc/programs/html (HTML files) and
share/EMBOSS/doc/programs/text (plain text, as used by the tfm program).
```

You can also use the EMBOSS application called **tfm** to display the plain text documentation of another application, e.g.:

```
tfm seqret
```

Q: Is there a quick guide?

A: There is a dated quick guide provided with the source code as the file doc/manuals/ emboss_qg.pdf. A Postscript file is also available. However, the primary source of up-to-date information is the website. The quick guide was not written by the EMBOSS team and it is therefore not maintained by them.

Q: Is there a table of substitutes for GCG programs?

A: A list of GCG and corresponding EMBOSS applications is available in the *EMBOSS User's Guide*:

```
http://emboss.open-bio.org/
```

B.1.10   Applications

Q: Plotting with **pepwheel** gives interesting output.

```
pepwheel -turns=8 -send=30 sw:p77837 -auto
```

gives a helical wheel plot but the residues are plotted so every two circles are sat on top of one another.

A: That is the correct answer. Instead of 3.6 residues per turn (5 turns in 18 steps), you seem to have a helix with 8 turns in 18 steps (4 in 9). Try

```
-turns=4 -steps=9
```

but only if you are sure that is the way your helix goes.

Q: In **prettyplot**, how do you specify an output file name for the plot file?

A: **prettyplot -auto ~/wordtest/globin-nogap.msf -graph ps** creates prettyplot.ps The name is generated automatically. To set this to something more descriptive use **-goutfile**, i.e.:

```
prettyplot -auto ~/wordtest/globin-nogap.msf -graph ps -goutfile=hello
```
Creates `hello.ps`

Q: I'm using the editor **mse**, but I don't know how to save my edited sequences at the end of the editing session.

A: Use `Control-Z` to get into command mode. Then use the **SAVE** command which will prompt for the file name.

B.1.11 EMBASSY

Q: What benefits do I gain from using the associated (EMBASSY) versions of software?

A: You can read any sequence type that EMBOSS can handle.

The associated software will use the EMBOSS ACD files so the naming of output/ input files is taken care of and will check all values before the program is run. Command line arguments are used instead interactive menu based ones.

B.1.12 ACD

Q: What is an ACD files?

A: Every EMBOSS and EMBASSY program has an ACD (AJAX command definition) file which describes the application, its options (parameters) and command line interface. The ACD file controls the behaviour of the application at the command line, particularly, all the user input operations. It includes an application definition (describing the application itself), a data definition for every application option (which describe the program parameters), attributes of the data definitions and the datatype of each option.

B.1.13 Interfaces

Q: What types of interface are available?

A: The EMBOSS command line is the simplest and best supported way to launch applications. The command line behaviour is consistent across all applications and provides a simple interface to (and complete control over) them. There are also web interfaces, graphical user interfaces, workflow and various other types of interface. These simplify, at least for some people, the ways to run EMBOSS applications.

B.1.14 Graphics

Q: What graphics options are available?

A: To see what graphics drivers are available type **?** at the prompt for the graph type and this will give you a list. Here are some of those:

ps (Postscript)

cps (Colour Postscript)

x11 (X display. Also called xterm and xwindows)

hpgl (HP Laserjet III, HPGL emulation mode)

png (PNG: you will need **libpng**, **zlib** and **gd** for this)

gif (GIF (you will need **libpng**, **zlib** and **gd** for this)

tek (Tektronix Terminal)

none (None)

data (Writes out points to a file for graphs)

meta (**plplot** meta file)

Q: What are the plotters and pen colours?

A: The hp drivers presume the pens are loaded as:

SP1 (black)

SP2 (white)

SP3 (red)

SP4 (green)

SP5 (blue)

SP6 (cyan)

SP7 (magenta)

SP8 (yellow)

Otherwise your output will have different colours.

Q: Browsing the GNU site I came across libplot, libxmi and plot utils. Would these be a suitable replacement for PLPLOT?

A: So far they are still GPL licensed rather than LGPL. Robert Maier promised a while back to LGPL them but has not yet. This is a pity as it makes all the difference for linking in third-party applications to EMBOSS.

Q: I get error messages when I try to display **X11** on my PC. I am running the Hummingbird Exceed X11 emulator.

A: This should only affect versions of EMBOSS pre-5.0.0. The Hummingbird Exceed X11 emulator (and maybe other systems) use the 'TrueColor' display by default. You should change the configuration settings so that it uses 'PseudoColor'. In v6.1 of Hummingbird Exceed, this is done by clicking on the Exceed section of the Windows 98 toolbar at the bottom of the screen. Select Tools, then Configuration, then Screen Definition. A window will appear. In the 'Server Visual' popup menu on the left, select PseudoColor. Click on OK.

B.1.15   Internals

Q: Is there a maximum size for sequences?

A: The maximum size for any program depends only on how much memory your system has. Of course, some programs (and some program options) can take up too much memory, or simply run very slowly. You might have a constraint imposed on your usage of memory. Try using the UNIX command **limit** to look at such constraints. Try using the UNIX command **unlimit** or **ulimit** to remove the constraints, e.g.:

```
unlimit stacksize
unlimit vmemoryuse
```

Q: GCG has a somewhat arbitrary fragment length limit of 2500 bp for **gel***. Is there a similar limit for **mse** under EMBOSS?

A: No. The **mse** package has no limit, you are only limited by how much memory you have.

Q: I am trying to write a web interface for an EMBOSS program and run Apache. The program complains that there is no HOME directory set.

A: Just put these at the top after your use CGI (whatever) statements.

```
$ENV[HOME] =web_owner_root i.e. /usr/local/apache
$ENV[EMBOSS_DATA] =emboss_data_directory
```

These two are important, but you can also pass other 'constants'.

B.1.16 Software development

Q: I am thinking of contributing software – how do I proceed?

A: Mail the EMBOSS developers at emboss-bug@emboss.open-bio.org

Index

Symbols

./configure command (*see* Configure script)
.embossrc file (*see* EMBOSS configuration files)

AAINDEX, 23, 120
Access methods (*see* Database)
ACD files, 149
  location, 32, 143
acdDirToParse, 86
aclocal, 136
AIX, 17, 76, 77
aixshadow, 17, 76, 77
AJAX command definition files (*see* ACD files)
Apache Axis, 66, 68
apache servers, 81
Apache Tomcat, 67, 68
API, xi
Application
  behaviour control using global qualifiers and
    environment variables, 27
  data files (*see* Data files)
  version number (*see* Version number)
Application ACD files (*see* ACD files)
Application documentation, 86, 147
Application version number (*see* Version number)
Arbitrary limits, xi
Attributes
  of databases (*see* Database)
Authentication, 71, 76
autoconf, 136
automake, 136
axis (*see* Apache Axis)

BBSRC, ix
Binaries, 8, 36, 66, 89, 136
BioPerl, 130
BLAST databases, 107, 113, 127
  Indexing and configuring, 127, 136
BLAST index files (*see* Database indexing)
Bugfix file (*see* README.fixes file)

C programming libraries, xi
CACHESIZE, 122
checkout command, 136
Codon usage table, 121
Compilation, 58, 136
  of EMBASSY (*see* EMBASSY packages)
  of EMBASSY pacakges, 56, 58
  of EMBASSY packages (*see* EMBASSY packages)
  of GD, 48
  of libharu, 50
  of LIBPNG, 47, 136

  of NCURSES, 51
  of ZLIB, 45, 136
  reporting compilation errors, 19, 59
  troubleshooting, 44
  using make, 18, 59
  using make clean, 12, 24, 44, 136
  using make install, 18, 20, 24, 34,
    44, 59, 136
  using make uninstall, 24, 60
Compilation tests, xii
config.status file, 8, 19, 56, 59
Configuration, 7, 26, 55, 136
  advanced options for, 12
  of EMBASSY packages (*see* EMBASSY packages)
  of EMBOSS to use graphics, 8
  options to avoid, 15
  site-wide, 27
  using configure script (*see* Configure script)
  with PDF graphics, 10
  with PNG graphics, 9
Configure script, 7, 12, 55, 136
CUTG, 121
CVS, 136

Data directory, 27, 32, 33, 86, 121
Data files, 22, 33, 119, 136, 143
Data formats
  of sequences (*see* Sequence formats)
Database, 92, 107, 119
  access methods, 93, 94, 98, 107, 108, 117
  attributes, 93, 98
  configuration (*see* Database configuration)
  definitions (*see* Database definitions)
  farms, 118
  flatfile (*see* Flatfile databases)
  query levels, 93
  relational (*see* Relational databases)
  search fields, 101
  sequence, 92, 107, 122
Database configuration, 92, 93, 129
  FASTA, 129
  Flatfile, 123
  GCG format, 126
Database definitions, 26, 92, 94, 98, 107
  example, 95
  testing, 97
  testing your database definitions, 124
Database indexing, 92, 122
  BLAST, 114, 127, 136
Debugging
  configuration for, 15
Dependencies, 15, 43, 44, 50, 89
Developers (CVS) release (*see* Release)

Printed in the United States
by Baker & Taylor Publisher Services